Math Success Practice Workbook Fractions 4th 5th 6th Grades

A Complete Guide to Master 2500+ Fractions Tasks and Word Problems with Test Prep

We create our workbooks with love and great care.
For any issues with your workbook, such as printing errors, typos, faulty binding, or something else, please do not hesitate to contact us at: info@homerunpress.com
We will make sure you get a replacement copy immediately.

THANK YOU!

Copyright © 2020 Home Run Press, LLC. All rights reserved.

Home Run Press, LLC 1603 Capitol Ave. Suite 310 A551 Cheyenne, WY 82001 USA

The purchase of this material entitles the buyer to reproduce worksheets and activities for classroom use only – not for commercial resale. Reproduction of these materials for an entire school or district is strictly prohibited. No part of this book may be reproduced (except as noted before), stored in retrieval system, or transmitted in any form or by any means (mechanically, electronically, photocopying, recording, etc.) without the prior written consent of Home Run Press, LLC.

First published in the USA 2020.   ISBN 9781952368257

Table of Contents

Introduction to Fractions — 8

Multiplying Fractions — 9

Practice: Multiplying Fractions. Sketching Fractions — 10

Reducing Fractions to Simplest Form. The Common Fractor — 11

Practice: Reducing Fractions to Simplest Form Using Fractions Strips — 12

Practice: Equivalent Fractions in Higher Terms — 13

Sketching Fractions — 14

Practice: Word Problem. Sketching Fractions — 15

Practice: Equivalent Fractions with the Given Numerator — 16

Practice: Equivalent Fractions with the Given Denominator. Comparing Fractions — 17

Practice: Equivalent Fractions with the Given Numerator. Sketching Fractions — 18

Practice: Equivalent Fractions with the Given Denominator. Sketching fractions — 19

Reducing Fractions to Simplest Form. The greatest Common Factor — 20

Equivalent Fractions in Lower Terms. Common Factors — 21

Practice: Reducing Fractions to Simplest Form. Comparing Fractions — 22

Practice: Reducing Fractions to Simplest Form. GCF — 23

Practice: Reducing Fractions to Simplest Form. GCF — 24

Practice: Reducing Fractions to Simplest Form. GCF. Comparing — 25

Practice: Reducing Fractions to Simplest Form. GCF. Comparing — 26

Practice: Reducing Fractions to Simplest Form. GCF. Comparing. — 27

Practice: Prime Factors — 28

Reducing Fractions to Simplest Form Using the Divide-Numerators-and-Denominators-Approach — 29

Reducing Fractions to Lower Terms — 30

Practice: Reducing Fractions to Simplest Form — 31

Table of Contents

Reducing Fractions to Simplest Form Using the Divide-Numerators-and-Denominators-Approach — 32

Reducing Fractions to Lower Terms — 33

Practice: Reducing Fractions to Simplest Form. Minecraft Factor Tree — 33

Practice: Prime Factors. Minecraft Factor Tree — 34

Practice: Prime Factors. Minecraft Factor Tree — 35

Practice: Prime Factors. Minecraft Factor Tree. Coloring — 36

Practice: Prime Factors. Minecraft Factor Tree — 37

Word Search. Maze — 38

Comparing fractions with Unlike Denominators — 39

Comparing fractions with Unlike Denominators — 40

Practice: Comparing Fractions with Unlike Denominators. Coloring — 41

Practice: Comparing Fractions with Unlike Denominators. Coloring — 42

Word Problems: Finding a Part of a Whole — 43

Word Problems: Finding a Part of a Whole. Coloring — 44

Practice: Word Problems: Finding a Part of a Whole — 45

Practice: Reducing Fractions to Simplest Form. Comparing fractions. Word Problem: Finding the Area. Test Prep — 46

Practice: Prime Factors. Minecraft Factor Tree — 47

Word Search. Coloring — 48

Practice: Reducing Fractions to Simplest Form. Comparing fractions — 49

Practice: Prime Factors. Minecraft Factor Tree — 50

Practice: Equivalent Fractions. Common Factors. Word Problem. Test Prep — 51

Practice: Word Problems: Finding a Part of a Whole — 52

Practice: Reducing Fractions to Simplest Form. Comparing fractions. Word Problem: Finding the Area. Test Prep — 53

Practice: Word Problems: Finding a Part of a Whole. Test Prep — 54

Table of Contents

Practice: Reducing Fractions to Simplest Form. Comparing fractions. Word Problem: Finding the Area. Test Prep. Coloring	55
Word Search. Coloring	56
Comparing Fractions with Unlike Numerators and Denominators	57
Comparing Fractions with Unlike Numerators and Denominators	58
The Least Common Denominator	59
The Least Common Denominator	60
Practice: Comparing fractions. Finding LCD	61
Practice: Comparing fractions. Finding LCD. Word Problem	62
Practice: Comparing fractions. Finding LCD. Word Problem	63
Practice: Comparing fractions. Finding LCD. Word Problem. Coloring	64
Word search. Coloring	65
Practice: Comparing fractions. Finding LCD. Word Problem. Coloring	66
Comparing Fractions with Unlike Numerators and Denominators Using a Number Line	67
Comparing Fractions with Unlike Numerators and Denominators Using a Number Line	68
Practice: Comparing fractions. Word Problem. Coloring	69
Word search. Coloring	70
Identifying Proper and Improper Fractions	71
Mixed Numbers. Practice: Identifying Proper and Improper Fractions	72
Practice: Identifying Proper and Improper Fractions Word Problem. Coloring	73
Converting Improper Fractions to Mixed numbers	74
Practice: Converting Improper Fractions to Mixed Numbers. Word problem	75
Practice: Converting Improper Fractions to Mixed Numbers. Word problems	76
Maze. Word search	77
Converting Mixed Numbers to Improper Fractions	78

Table of Contents

Practice: Converting Mixed Numbers to Improper Fractions. Word problem	79
Practice: Converting Mixed Numbers to Improper Fractions. Word problems	80
Practice: Converting Improper Fractions to Mixed Numbers. Word problems. TestPrep	81
Practice: Converting Mixed Numbers to Improper Fractions. Word problems. Test Prep	82
Word search. Coloring	83
Adding with Like and Unlike Denominators	84
Adding Fractions with Unlike Denominators	85
Adding Fractions with Unlike Denominators	86
Adding Fractions with Unlike Denominators	87
Adding Fractions with Unlike Denominators. Coloring	88
Practice: Adding Fractions with Unlike Denominators. Test Prep	89
Practice: Adding Fractions with Unlike Denominators. Test Prep	90
Practice: Adding Fractions with Unlike Denominators. Test Prep	91
Practice: Adding Fractions with Unlike Denominators. Test Prep. Coloring	92
Practice: Adding Fractions with Unlike Denominators. Test Prep	93
Maze. Word search. Coloring	94
Subtracting Fractions with Unlike Denominators	95
Subtracting Fractions with Unlike Denominators	96
Subtracting Fractions with Unlike Denominators	97
Subtracting Fractions with Unlike Denominators	98
Practice: Subtracting Fractions with Unlike Denominators. Word problem	99
Practice: Subtracting Fractions with Unlike Denominators. Word problem	100
Practice: Subtracting Fractions with Unlike Denominators. Word problem	101
Practice: Adding and Subtracting Fractions. Maze. Coloring	102

Table of Contents

Practice: Subtracting Fractions with Unlike Denominators. Word problem	103
Practice: Subtracting Fractions. Word problem. Coloring	104
Word search. Coloring	105
Practice: Adding and Subtracting Fractions. Maze. Coloring	106
Word search. Coloring	107
Subtracting Whole Numbers and Fractions	108
Practice: Subtracting Whole Numbers and Fractions. Finding a Part of a Whole. Test Prep. Coloring	109
Adding Mixed Numbers	110
Adding Mixed Numbers	111
Practice: Adding Mixed Numbers. Test Prep	112
Subtracting Mixed Numbers	113
Subtracting Mixed Numbers	114
Practice: Subtracting Mixed Numbers. Test Prep	115
Multiplying Mixed Numbers. Practice: Multiplying Mixed Numbers. Test Prep	116
Word search. Coloring	117
Practice: Multiplying Mixed Numbers. Test Prep	118
Practice: Adding and Subtracting Fractions. Maze. Coloring	119
Dividing a Whole Number by a Fraction	120
Dividing a Fraction by a Whole Number	121
Dividing a Fraction by a Whole Number	122
Dividing a Fraction by a Whole Number Using a Reciprocal	123
Practice: Dividing a Fraction by a Whole Number. Word Problem. Coloring	124
Practice: Dividing a Fraction by a Whole Number. Word Problem. Coloring	125
Practice: Dividing a Fraction by a Whole Number. Word Problem	126

Table of Contents

Coloring. Word Search	127
Dividing a Whole Number by a Fraction	128
Practice: Dividing a Whole Number by a Fraction. Word Problem. Coloring	129
Practice: Dividing a Whole Number by a Fraction. Word Problem. Coloring	130
Practice: Dividing or Multiplying Fractions. Word Problem	131
Dividing Proper Fractions	132
Practice: Dividing Fractions. Word Problem. Coloring	133
Practice: Dividing Fractions. Word Problem. Coloring	134
Practice: Dividing Fractionsand Mixed Numbers. Word Problem. Coloring	135
Maze. Word Search. Coloring	136
Answers	137

Kids learn and stay engaged, thanks to puzzles, coloring, mazes, word search tasks, along with challenging math problems. These methods help kids understand math concepts, master math skills, and even help struggling students gain the confidence to improve their math comprehension and testing.

These activities are perfect for daily practice, morning work, homework, math centers, early finishers, test preparation, assessment, or for high school students struggling with fractions.

They will work great in 4th Grade to challenge students. They are perfect in 5th-6th Grades to practice fractions, but they also might work in 7th-9th Grades as a review or for those who are struggling with math.

Hi. I'm Sunny. For me, everything is an adventure. I am ready to try anything, take chances, see what happens - and help you try, too! I like to think I'm confident, caring and have an open mind. I will cheer for your success and encourage everyone! I'm ready to be a really good friend!

I've got a problem. Well, I've always got a problem. And I don't like it. It makes me cranky, and grumpy, impatient and the truth is, I got a bad attitude. There. I said it. I admit it. And the reason I feel this way? Math! I don't get it and it bums me out. Grrrr!

Not trying to brag, but I am the smartest Brainer that ever lived - and I'm a brilliant shade of blue. That's why they call me Smarty. I love to solve problems and I'm always happy to explain how things work - to help any Brainer out there! To me, work is fun, and math is a blast!

I scare easily. Like, even just a little ...Boo! Oh wow, I've scared myself! Anyway, they call me Pickles because I turn a little green when I get panicky. Especially with new stuff. Eek! And big complicated problems. Really any problem. Eek! There, I did it again.

Hi! Name's Pepper. I have what you call a positive outlook. I just think being alive is exciting! And you know something? By being friendly, kind and maybe even wise, you can have a pretty awesome day every day on this amazing planet.

A famous movie star once said, "I want to be alone." Well, I do too! I'm best when I'm dreaming, thinking, and in my own world. And so, I resist! Yes, I resist anything new, and only do things my way or quit. The rest of the Brainers have math, but I'd rather have a headache and complain. Or pout.

1. <u>Read and write</u> the missing numbers.

$\frac{1}{1}$ $\frac{2}{2}$ $\frac{4}{4}$ $\boxed{1=\frac{1}{1}}$

The **numerator** is the numeral that is written above the bar. → $\frac{8}{8}$

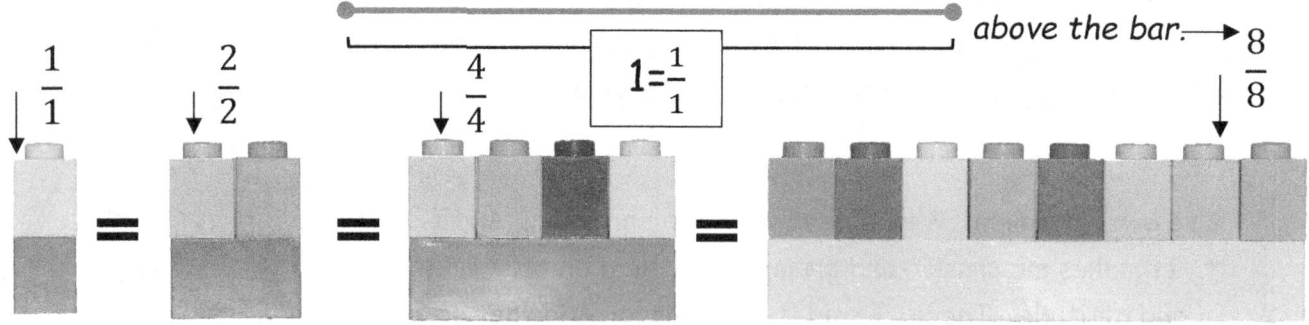

The numerator represents how many equal PARTS are being identified:

The **denominator** is the numeral that is written under the bar.

$$\text{Whole} = 1 = \frac{2}{2} = \frac{4}{4} = \frac{8}{8}$$

The denominator represents the number of equal parts in the WHOLE (the whole is divided into the two, four, or eight equal parts)

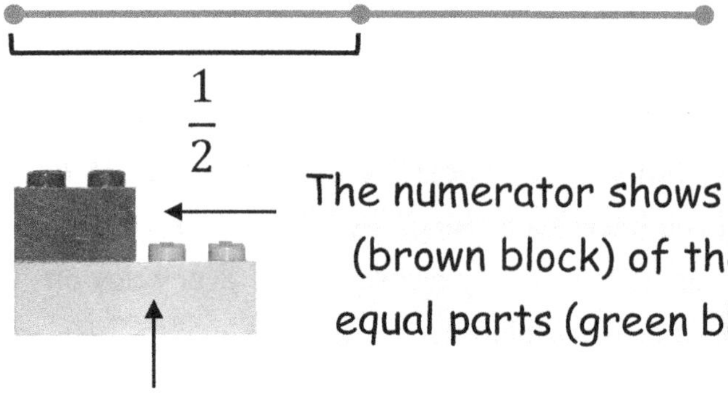

$\frac{1}{2}$

The **numerator** 1 tells that 1 part of the whole is being counted.

A half = $\frac{1}{2}$ ← The **denominator** 2 tells that a whole was divided into 2 equal parts.

The numerator shows you one (brown block) of the two equal parts (green blocks).

The denominator shows you the number of the equal parts (green blocks) in the whole.

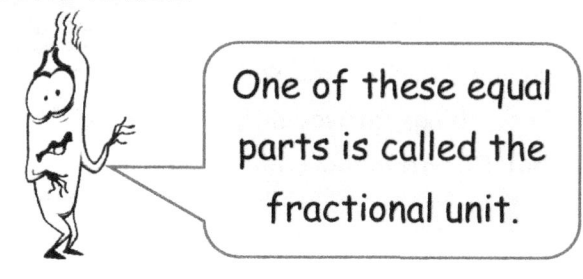

One of these equal parts is called the fractional unit.

The fractional unit for this block is $\frac{1}{4}$

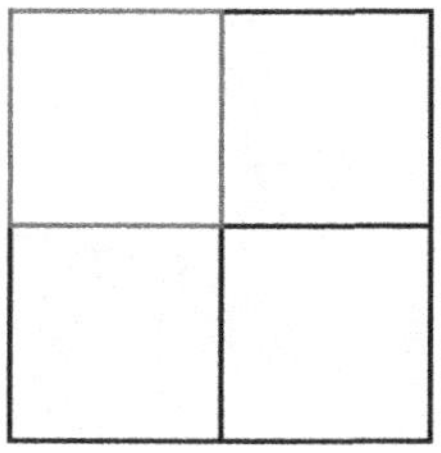

Aha... I start with multiplying the fractions, right? So, to multiply the fractions I need to multiply out the numerators and then, I need to multiply out the denominators.

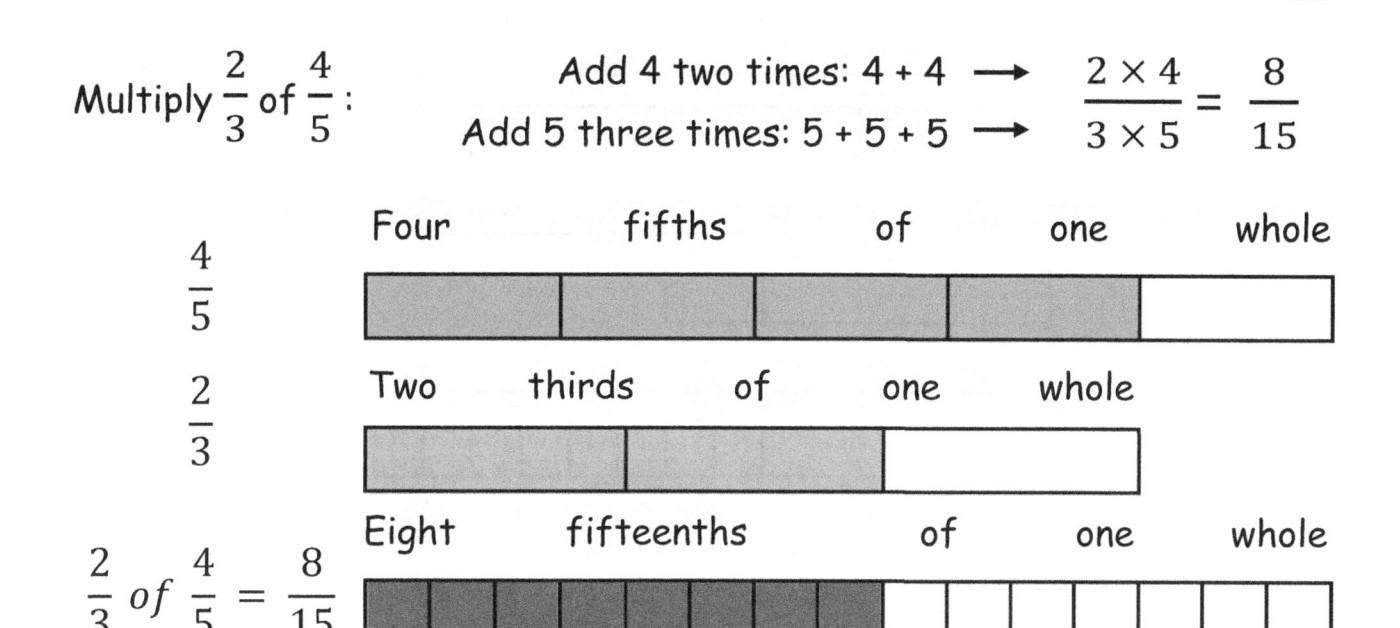

Multiply $\frac{2}{3}$ of $\frac{4}{5}$:   Add 4 two times: 4 + 4 ⟶ $\frac{2 \times 4}{3 \times 5} = \frac{8}{15}$
Add 5 three times: 5 + 5 + 5 ⟶

$\frac{4}{5}$ Four   fifths   of   one   whole

$\frac{2}{3}$ Two   thirds   of   one   whole

$\frac{2}{3}$ of $\frac{4}{5} = \frac{8}{15}$ Eight   fifteenths   of   one   whole

Multiply $\frac{3}{5}$ of $\frac{1}{4}$: $\frac{add\ 1\ three\ times: 1+1+1}{add\ 4\ five\ times: 4+4+4+4+4} = \frac{3 \times 1}{5 \times 4} = \frac{3}{20}$

$\frac{1}{4}$ One   unit   of   one   fourth

$\frac{3}{5}$ Three units of one fifth

$\frac{3}{5}$ of $\frac{1}{4} = \frac{3}{20}$ Three   units   of   one   twentieth

Multiplying is easy!   **Yes, a common denominator is not needed!**

1. <u>Multiply</u> fractions. <u>Color</u> the fractions bars. <u>Write</u> the missing numbers and words.

$$\frac{2 \times 2}{4 \times 5} = \frac{add\ \underline{}\ \underline{}\ times:\underline{}}{add\ \underline{}\ \underline{}\ times:\underline{}} = \underline{}$$

—
—

Two _____ of one whole

Two _____ of one whole

— of — =   _____ of one whole

= —

2. <u>Multiply</u> fractions. <u>Draw and color</u> the fractions bars. <u>Write</u> the missing numbers.

$$\frac{5 \times 4}{6 \times 5} = \frac{add\ \underline{}\ \underline{}\ times:\underline{}}{add\ \underline{}\ \underline{}\ times:\underline{}} = \underline{}$$

—
—

Four units of _____

Five units of _____

— of — =   _____ units of _____

= —

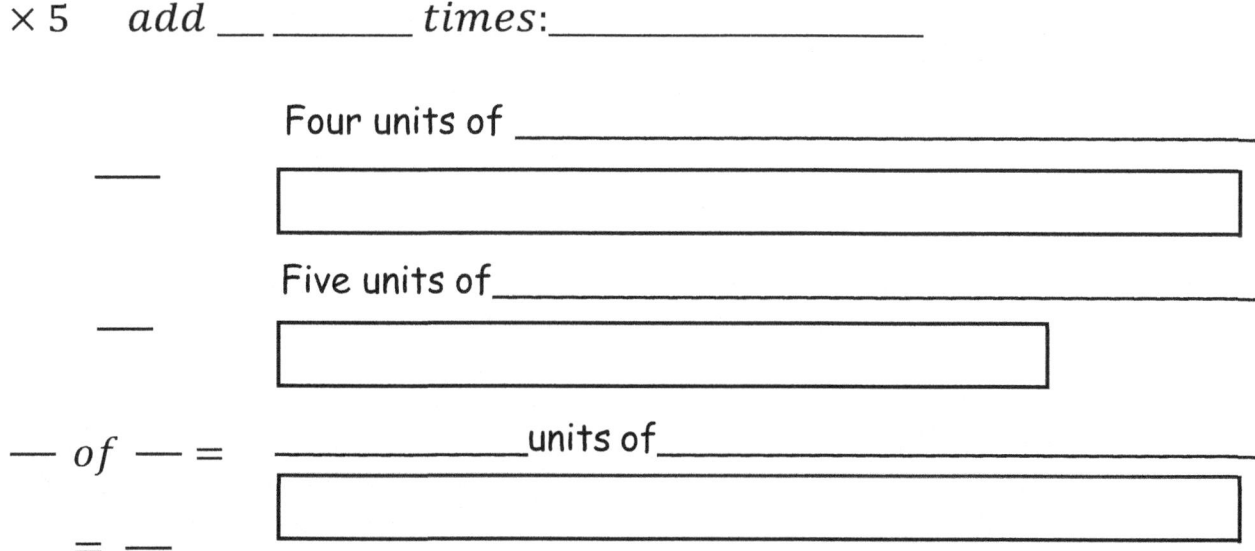

1. Equivalent fractions. <u>Answer</u> the questions and <u>write</u> the missing numbers.

Equivalent fractions? Oh, boy! What's that?

How can I explain? Equivalent fractions show the same value of the whole.

I circled one third black. The whole is divided into 3 equal parts.

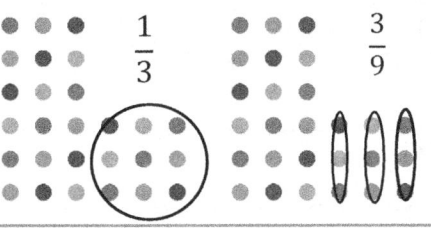

$\frac{1}{3}$ $\frac{3}{9}$

I circled three ninths black. The whole is divided into 9 equal parts. I circled 9 dots in both pictures.

The numerators and denominators are different, but the value is the same. Take a half of the block: $\frac{1}{2}$ (a half). If you multiply the numerator and the denominator BY THE SAME NUMBER called the COMMON FACTOR, say, 2, you'll get a different fraction - $\frac{2}{4}$ (two fourths), but their value will stay the same:

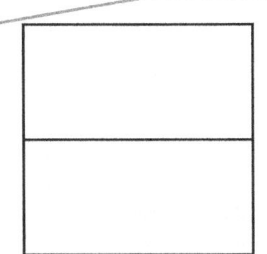

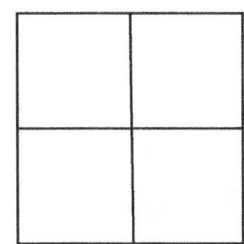

2 is the *common factor*

$$\frac{1 \times 2}{2 \times 2} = \frac{2}{4}$$

2 is the *common factor*

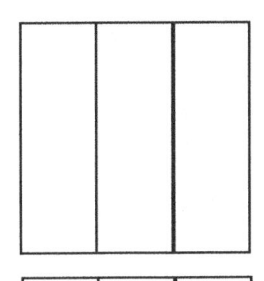

To find an equivalent fraction of $\frac{1}{3}$ I need to multiply the numerator and the denominator BY THE SAME NUMBER called the COMMON FACTOR (say, 3): $\frac{1 \times 3}{3 \times 3} = \frac{3}{9}$.

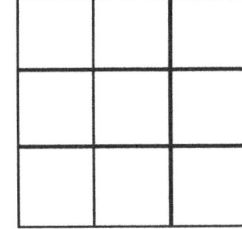

3 is the *common factor*

$$\frac{1 \times 3}{3 \times 3} = \frac{3}{9}$$

3 is the *common factor*

$\frac{1}{3}$ and $\frac{3}{9}$ represent the same part of the whole

1. <u>Fill in</u> the missing numbers using the fractions strips.

$$\frac{1\times 2}{2\times 2} = \frac{}{}$$

$$\frac{1\times 4}{2\times 4} = \frac{}{}$$

$$\frac{1\times 2}{3\times 2} = \frac{}{}$$

$$\frac{1\times 2}{5\times 2} = \frac{}{}$$

$$\frac{1\times 2}{4\times 2} = \frac{}{}$$

$$\frac{1\times}{3\times 2} = \frac{}{} \qquad \frac{1\times}{5\times 2} = \frac{}{} \qquad \frac{1\times}{4\times 2} = \frac{}{}$$

$$\frac{1\times}{2\times 5} = \frac{}{} \qquad \frac{1\times}{3\times 3} = \frac{}{} \qquad \frac{1\times}{2\times 4} = \frac{}{}$$

$$\frac{1\times}{\times 2} = \frac{}{8} \qquad \frac{1\times}{\times 3} = \frac{}{6} \qquad \frac{1\times}{\times 5} = \frac{}{10}$$

$$\frac{1\times}{\times 4} = \frac{}{8} \qquad \frac{1\times}{\times 3} = \frac{}{9} \qquad \frac{1\times}{\times 5} = \frac{}{5}$$

$$\frac{1\times}{\times 6} = \frac{}{12} \qquad \frac{1\times}{\times 5} = \frac{}{15} \qquad \frac{1\times}{\times 4} = \frac{}{12}$$

1. <u>Find</u> an equivalent fraction with the given denominator in higher terms and <u>color</u> the fraction of each shape.

Red   Blue   Green   Yellow   Grey   Orange   Purple   Pink

$$\frac{1}{2} = \frac{}{4} = \frac{}{8} = \frac{}{10} = \frac{}{14} = \frac{}{16} = \frac{}{20} = \frac{}{40}$$

*How to create an **equivalent fraction** in higher terms?*

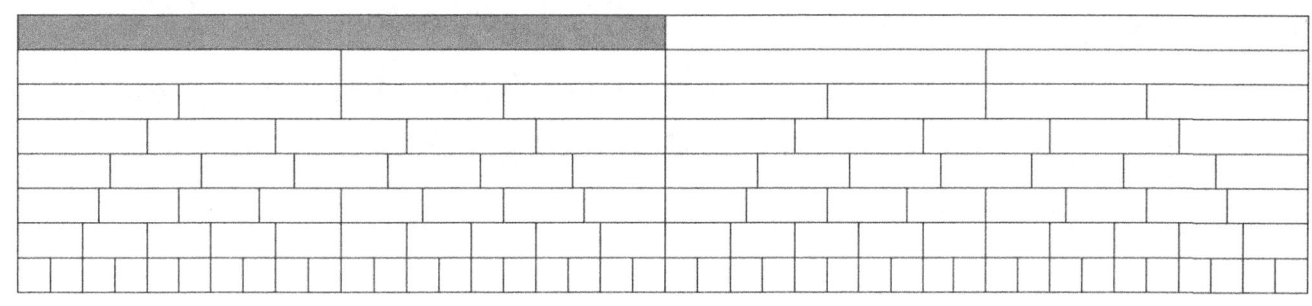

Red   Blue   Green   Yellow   Grey   Brown   Purple   Pink

$$\frac{1}{3} = \frac{}{6} = \frac{}{9} = \frac{}{12} = \frac{}{18} = \frac{}{21} = \frac{}{24} = \frac{}{36}$$

*! Multiply both the **numerator** and the **denominator** by the **same whole number**.*

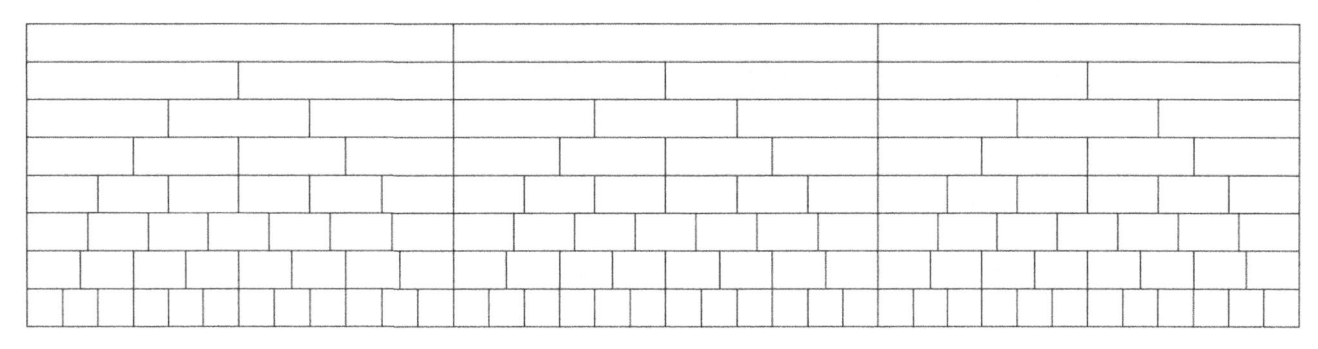

Red   Blue   Green   Yellow   Grey   Brown   Orange   Purple   White   Pink

$$\frac{1}{4} = \frac{}{8} = \frac{}{12} = \frac{}{16} = \frac{}{20} = \frac{}{24} = \frac{}{28} = \frac{}{32} = \frac{}{36} = \frac{}{40}$$

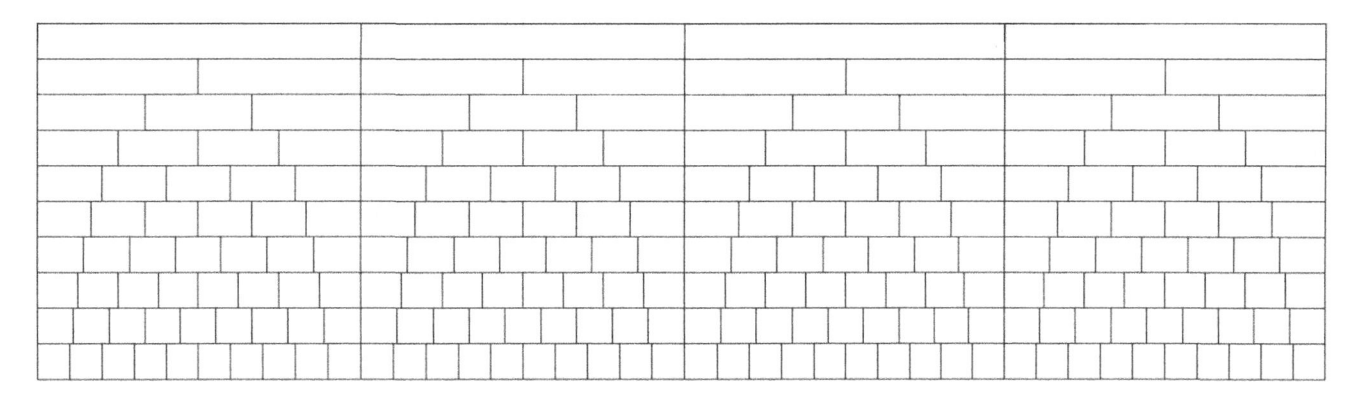

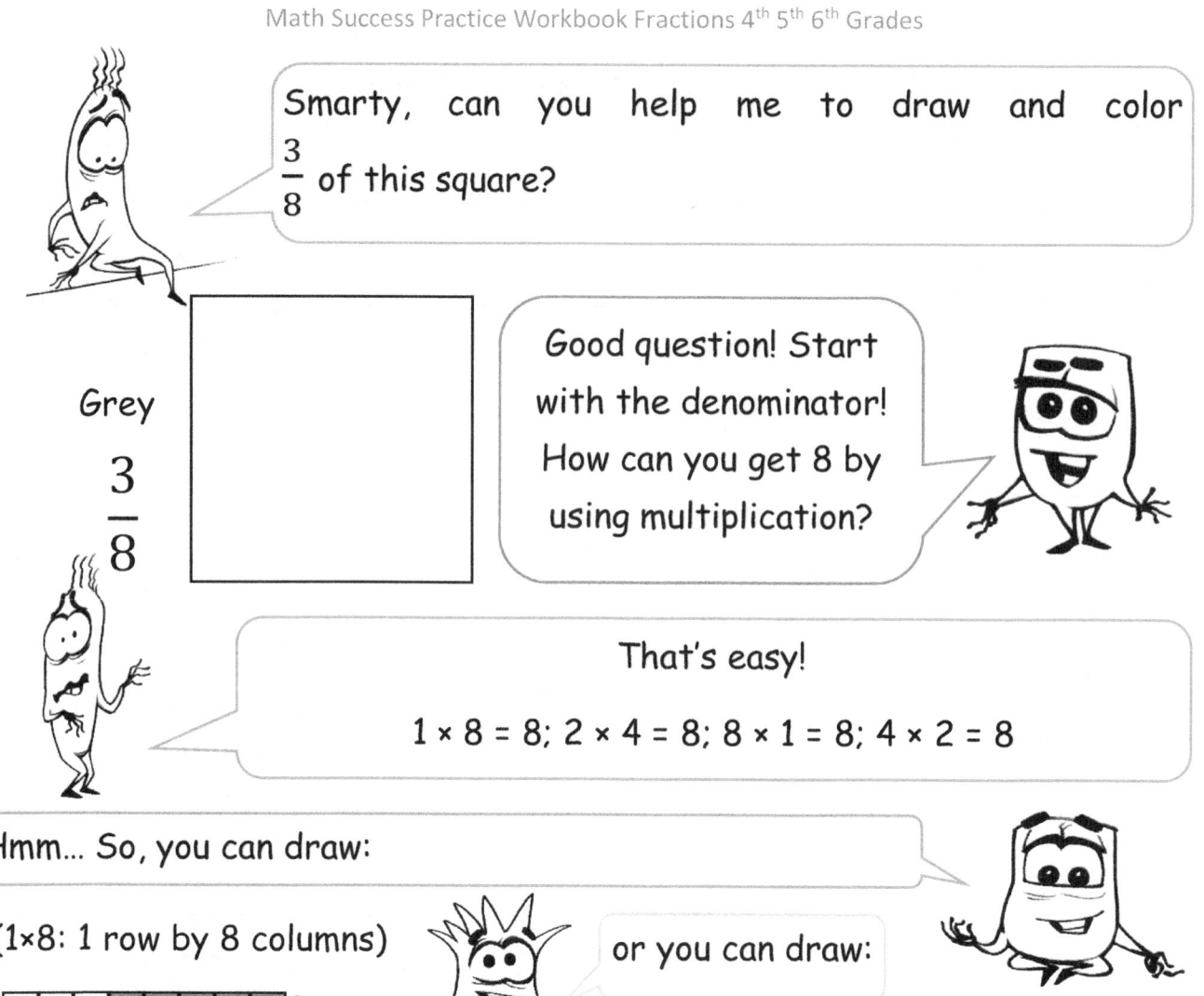

Math Success Practice Workbook Fractions 4th 5th 6th Grades

Smarty, can you help me to draw and color $\frac{3}{8}$ of this square?

Grey $\frac{3}{8}$

Good question! Start with the denominator! How can you get 8 by using multiplication?

That's easy!

1 × 8 = 8; 2 × 4 = 8; 8 × 1 = 8; 4 × 2 = 8

Hmm… So, you can draw:

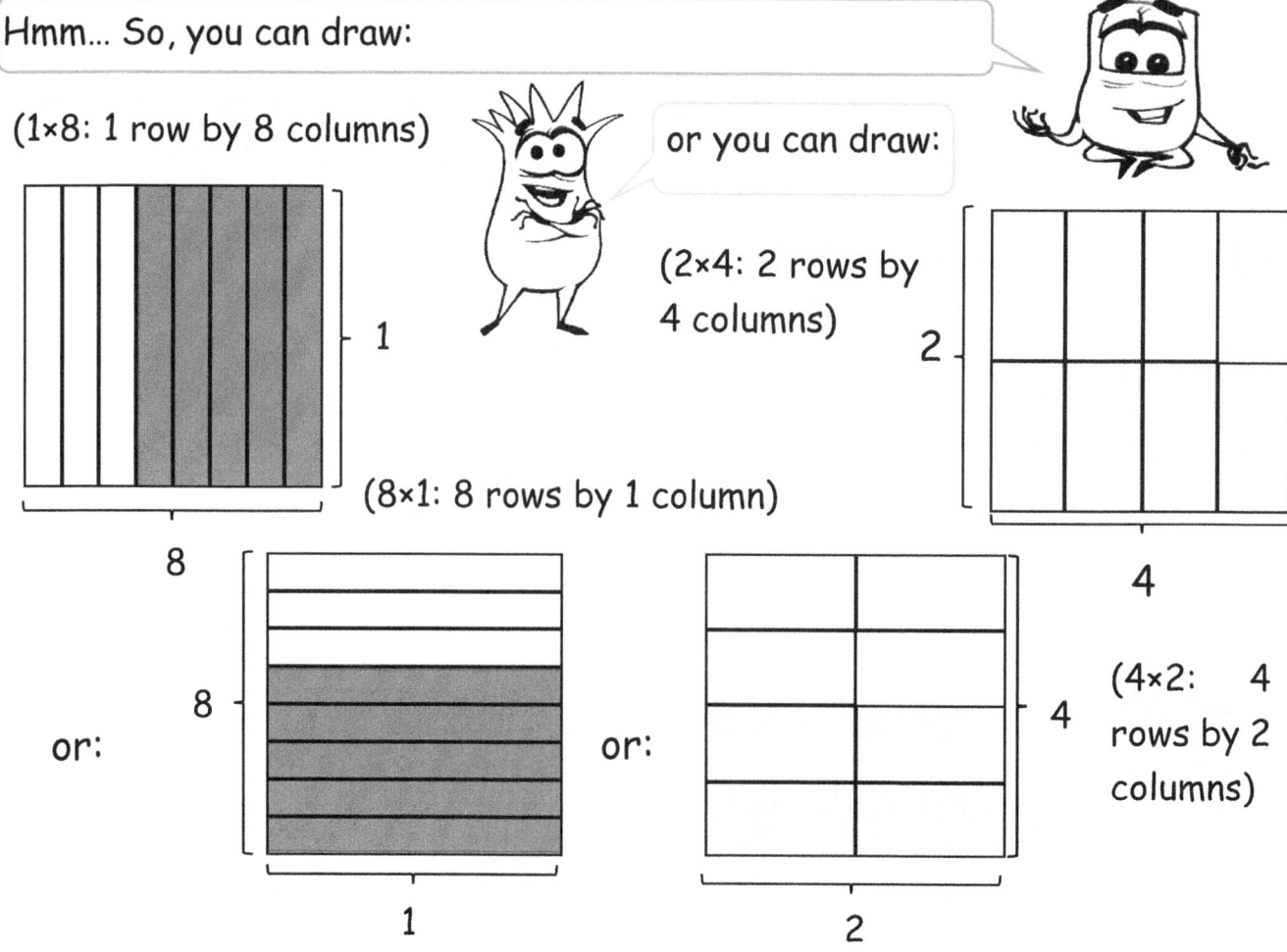

(1×8: 1 row by 8 columns)

or you can draw:

(2×4: 2 rows by 4 columns)

or:

(8×1: 8 rows by 1 column)

or:

(4×2: 4 rows by 2 columns)

1. Draw and color the fraction of each shape.

$\frac{4}{6}$

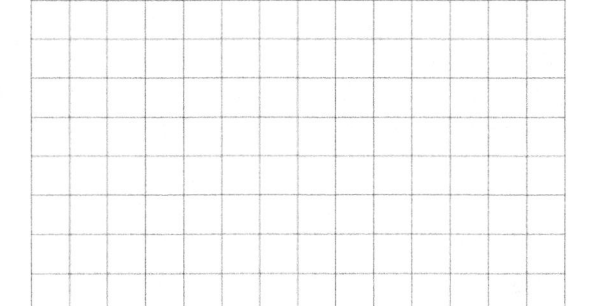

$\frac{6}{9}$

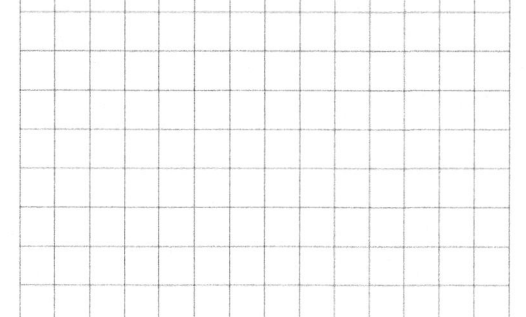

$\frac{3}{5}$

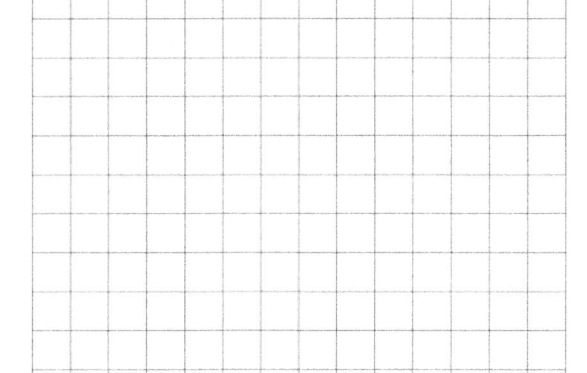

$\frac{1}{4}$

$\frac{2}{7}$

$\frac{5}{12}$

2. I had 26 blocks. I colored a $\frac{1}{2}$ of them. How many blocks were left white?

Answer: _____.

1. <u>Find</u> an equivalent fraction with the given numerator.

Step 1: Use the given fraction ($\frac{1}{2}$).

Step 2: The numerator 1 has been multiplied by 2 to get 2.

Step 3: To get an equivalent fraction, the denominator 2 must also be multiplied by 2 to get 4. Therefore $\frac{1}{2} = \frac{2}{4}$.

$$\frac{1}{2} = \frac{2}{4} = \frac{11}{\_\_} = \frac{4}{\_\_} = \frac{9}{\_\_} = \frac{6}{\_\_} = \frac{10}{\_\_} = \frac{8}{\_\_} = \frac{5}{\_\_} = \frac{7}{\_\_} = \frac{3}{\_\_} = \frac{12}{\_\_}$$

$$\frac{1}{3} = \frac{2}{6} = \frac{12}{\_\_} = \frac{4}{\_\_} = \frac{8}{\_\_} = \frac{11}{\_\_} = \frac{7}{\_\_} = \frac{5}{\_\_} = \frac{9}{\_\_} = \frac{10}{\_\_} = \frac{6}{\_\_} = \frac{3}{\_\_}$$

$$\frac{1}{4} = \frac{9}{\_\_} = \frac{3}{\_\_} = \frac{12}{\_\_} = \frac{5}{\_\_} = \frac{8}{\_\_} = \frac{11}{\_\_} = \frac{6}{\_\_} = \frac{2}{\_\_} = \frac{10}{\_\_} = \frac{7}{\_\_} = \frac{4}{\_\_}$$

$$\frac{1}{5} = \frac{8}{\_\_} = \frac{3}{\_\_} = \frac{12}{\_\_} = \frac{2}{\_\_} = \frac{10}{\_\_} = \frac{4}{\_\_} = \frac{5}{\_\_} = \frac{9}{\_\_} = \frac{6}{\_\_} = \frac{11}{\_\_} = \frac{7}{\_\_}$$

2. <u>Write</u> the fraction name under each picture.

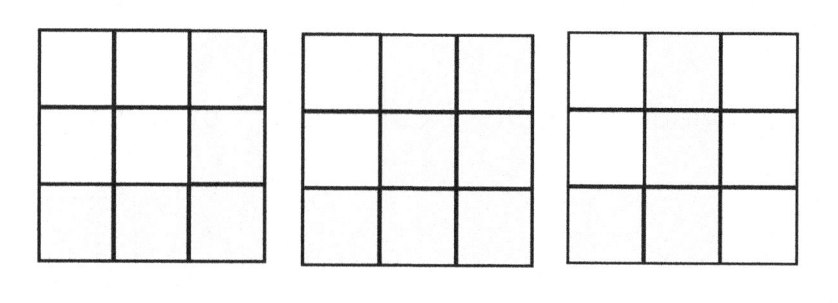

$$\frac{white\ blocks}{blocks\ in\ total}$$

$\frac{...}{...}$ $\frac{...}{...}$ $\frac{...}{...}$

1. <u>Find</u> an equivalent fraction with the given denominator.

Step 1: Use the given fraction ($\frac{1}{2}$).

Step 2: The denominator 2 has been multiplied by 9 to get 18.

Step 3: To get an equivalent fraction, the numerator 1 must also be multiplied by 9 to get 9. Therefore $\frac{1}{2} = \frac{9}{18}$.

$$\frac{1}{2} = \frac{9}{18} = \frac{}{4} = \frac{}{12} = \frac{}{22} = \frac{}{16} = \frac{}{8} = \frac{}{28} = \frac{}{40} = \frac{}{6} = \frac{}{10} = \frac{}{34}$$

$$\frac{1}{3} = \frac{7}{21} = \frac{}{9} = \frac{}{60} = \frac{}{15} = \frac{}{27} = \frac{}{12} = \frac{}{6} = \frac{}{18} = \frac{}{36} = \frac{}{48} = \frac{}{24}$$

$$\frac{1}{4} = \frac{}{24} = \frac{}{40} = \frac{}{16} = \frac{}{32} = \frac{}{12} = \frac{}{48} = \frac{}{8} = \frac{}{36} = \frac{}{20} = \frac{}{28} = \frac{}{56}$$

$$\frac{1}{5} = \frac{}{35} = \frac{}{15} = \frac{}{45} = \frac{}{20} = \frac{}{30} = \frac{}{65} = \frac{}{25} = \frac{}{70} = \frac{}{50} = \frac{}{10} = \frac{}{40}$$

2. <u>Which</u> fraction is bigger ($\frac{white\ blocks}{blocks\ in\ total}$)?

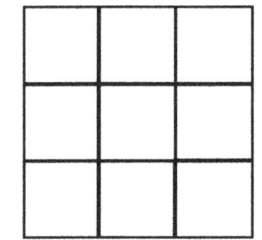

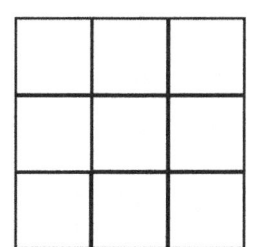

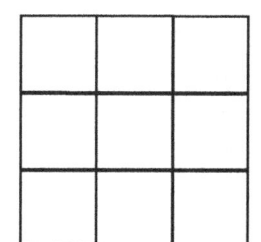

$\frac{}{} < \frac{}{} < \frac{}{}$

1. <u>Find</u> an equivalent fraction with the given numerator.

Step 1: Use the given fraction ($\frac{1}{6}$).
Step 2: The numerator 1 has been multiplied by 2 to get 2.
Step 3: To get an equivalent fraction, the denominator 6 must also be multiplied by 2 to get 12.
Therefore $\frac{1}{6} = \frac{2}{12}$.

$$\frac{1}{6} = \frac{2}{12} = \frac{9}{-} = \frac{4}{-} = \frac{12}{-} = \frac{6}{-} = \frac{10}{-} = \frac{8}{-} = \frac{3}{-} = \frac{7}{-} = \frac{11}{-} = \frac{5}{-}$$

$$\frac{1}{7} = \frac{2}{-} = \frac{7}{-} = \frac{12}{-} = \frac{5}{-} = \frac{9}{-} = \frac{3}{-} = \frac{8}{-} = \frac{6}{-} = \frac{10}{-} = \frac{11}{-} = \frac{4}{-}$$

$$\frac{1}{8} = \frac{2}{-} = \frac{6}{-} = \frac{4}{-} = \frac{9}{-} = \frac{3}{-} = \frac{11}{-} = \frac{8}{-} = \frac{5}{-} = \frac{10}{-} = \frac{7}{-} = \frac{12}{-}$$

$$\frac{1}{9} = \frac{5}{-} = \frac{3}{-} = \frac{9}{-} = \frac{2}{-} = \frac{11}{-} = \frac{7}{-} = \frac{12}{-} = \frac{4}{-} = \frac{10}{-} = \frac{6}{-} = \frac{8}{-}$$

2. <u>Draw</u> and <u>color</u> the fraction of each shape.

(2 × 2)   (2 × 4)   (4 × 3)

[ ]   [ ]   [ ]

$\frac{3}{4}$ $\frac{6}{8}$ $\frac{9}{12}$

1. <u>Find</u> an equivalent fraction with the given denominator.

Step 1: Use the given fraction ($\frac{1}{6}$).

Step 2: The denominator 6 has been multiplied by 3 to get 18.
Step 3: To get an equivalent fraction, the numerator 1 must also be multiplied by 3 to get 3.
Therefore $\frac{1}{6} = \frac{3}{18}$.

$$\frac{1}{6} = \frac{3}{18} = \frac{\_}{36} = \frac{\_}{12} = \frac{\_}{48} = \frac{\_}{66} = \frac{\_}{24} = \frac{\_}{78} = \frac{\_}{42} = \frac{\_}{60} = \frac{\_}{30} = \frac{\_}{54}$$

$$\frac{1}{7} = \frac{\_}{21} = \frac{\_}{49} = \frac{\_}{70} = \frac{\_}{14} = \frac{\_}{77} = \frac{\_}{42} = \frac{\_}{28} = \frac{\_}{63} = \frac{\_}{35} = \frac{\_}{84} = \frac{\_}{56}$$

$$\frac{1}{8} = \frac{\_}{24} = \frac{\_}{40} = \frac{\_}{16} = \frac{\_}{88} = \frac{\_}{72} = \frac{\_}{48} = \frac{\_}{64} = \frac{\_}{80} = \frac{\_}{96} = \frac{\_}{32} = \frac{\_}{56}$$

$$\frac{1}{9} = \frac{\_}{36} = \frac{\_}{18} = \frac{\_}{72} = \frac{\_}{27} = \frac{\_}{90} = \frac{\_}{45} = \frac{\_}{63} = \frac{\_}{99} = \frac{\_}{54} = \frac{\_}{81} = \frac{\_}{117}$$

2. <u>Draw</u> and <u>color</u> the fraction of each shape.

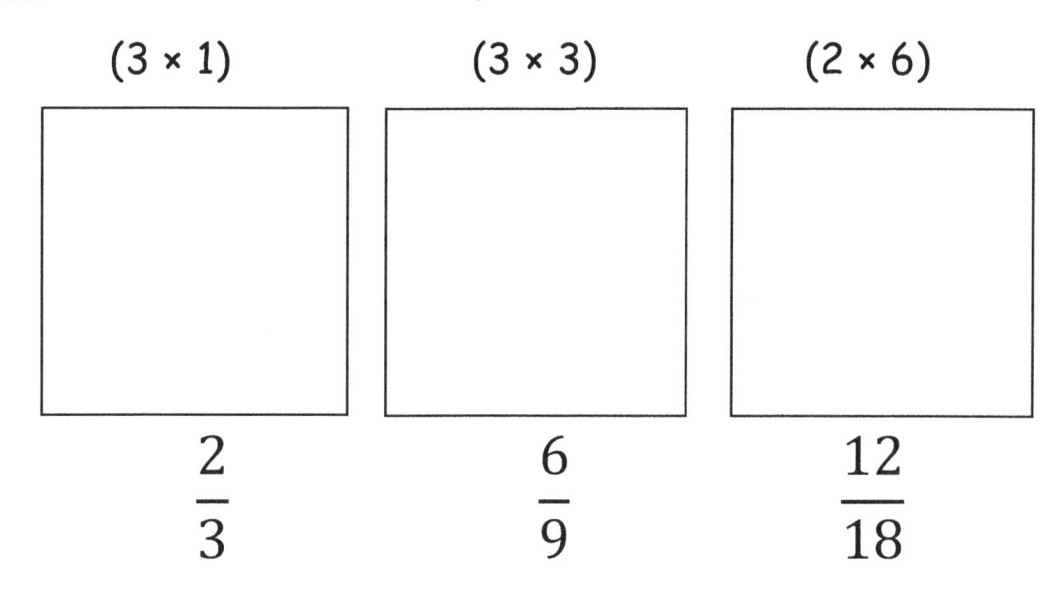

(3 × 1)   (3 × 3)   (2 × 6)

$\frac{2}{3}$ $\frac{6}{9}$ $\frac{12}{18}$

About simplifying... When you divide both the numerator and denominator by the same number, you simplify the fraction, or write it in lower terms, or reduce it to its lowest terms.

I colored white the same relative amount of my block ($\frac{1}{4}$). Though the sizes of the equal parts my blocks are divided into are different.

$$\frac{1}{4} = \frac{1 \times 1}{4 \times 1}$$

We divided $\frac{1}{4}$ by 2

$$\frac{2}{8} = \frac{1 \times 2}{4 \times 2}$$

We divided $\frac{1}{4}$ by 3

$$\frac{3}{12} = \frac{1 \times 3}{4 \times 3}$$

We divided $\frac{1}{4}$ by 4

$$\frac{4}{16} = \frac{1 \times 4}{4 \times 4}$$

All these fractions represent the same relative amount – one fourth.

The fraction $\frac{4}{16}$ is in higher terms, while $\frac{1}{4}$ is in its lowest terms. So $\frac{4}{16}$ was simplified or reduced to its lowest terms.

To simplify a fraction, divide both the numerator and the denominator by the same number and write a fraction in its simplest form:

$$\frac{2}{8} = \frac{(2 \div 2)}{(4 \div 2)} = \frac{1}{4}$$

$$\frac{3}{12} = \frac{(3 \div 3)}{(12 \div 3)} = \frac{1}{4}$$

$$\frac{4}{16} = \frac{(4 \div 4)}{(16 \div 4)} = \frac{1}{4}$$

Find the greatest factor both the numerator and the denominator can be divided evenly without any remainder.

The greatest common factor is 2
The greatest common factor is 3
The greatest common factor is 4

1. <u>Reduce</u> each fraction to simplest form.

Well, to reduce a fraction to its lowest form (or term) means that I need to divide the numerator (the top number) and the denominator (the bottom number) by the SAME GREATEST number.

$$\frac{(3 \div 3)}{(12 \div 3)} = \frac{1}{4}$$

$$\frac{3}{12} = \frac{(1 \times \cancel{3})}{(4 \times \cancel{3})} = \frac{1}{4}$$

I prefer to say that I cancel the 3s.

So, you reduce the fraction to make it look as simple as possible. What is the GREATEST number both 2 and 4 can be divided into? Right, they both can be divided by 2 (the **greatest common factor**): $\frac{2}{4} = \frac{(2 \div 2)}{(4 \div 2)} = \frac{1}{2}$.

To write an **equivalent fraction in lower terms**, divide both the numerator and denominator by the same number – **common factor (2)**.

$$\frac{2}{4} = \frac{1}{2}$$

$$\frac{(2 \div 2)}{(4 \div 2)} = \frac{1}{2}$$

I can divide both the numerator and the denominator by 2, 3, 4, 5, 6, and etc. I need to reduce $\frac{18}{42}$ to its lowest terms: $\frac{(18 \div 2)}{(42 \div 2)} = \frac{(9 \div 3)}{(21 \div 3)} = \frac{3}{7}$.

I don't change the VALUE of the fraction.

$$\frac{(16 \div 2)}{(40 \div 2)} = \frac{(8 \div 2)}{(20 \div 2)} = \frac{(4 \div 2)}{(10 \div 2)} = \frac{2}{5}$$

1. <u>Write</u> each fraction in its simplest form. <u>Write</u> the missing numbers.

$\dfrac{(3 \div 3)}{(6 \div 3)} = \dfrac{}{}$

The Greatest Common factor = 3

$\dfrac{(3 \div)}{(12 \div)} = \dfrac{}{}$

The Greatest Common factor = 3

$\dfrac{(2 \div)}{(14 \div)} = \dfrac{}{}$

The Greatest Common factor = ___

$\dfrac{(2 \div)}{(4 \div)} = \dfrac{}{}$

The Greatest Common factor = ___

$\dfrac{(5 \div)}{(20 \div)} = \dfrac{}{}$

The Greatest Common factor = ___

$\dfrac{(5 \div)}{(35 \div)} = \dfrac{}{}$

The Greatest Common factor = ___

$\dfrac{(4 \div)}{(24 \div)} = \dfrac{}{}$

The Greatest Common factor = ___

$\dfrac{(4 \div)}{(32 \div)} = \dfrac{}{}$

The Greatest Common factor = ___

$\dfrac{(6 \div)}{(18 \div)} = \dfrac{}{}$

The Greatest Common factor = ___

$\dfrac{(6 \div)}{(42 \div)} = \dfrac{}{}$

The Greatest Common factor = ___

$\dfrac{(7 \div)}{(14 \div)} = \dfrac{}{}$

The Greatest Common factor = ___

$\dfrac{(7 \div)}{(28 \div)} = \dfrac{}{}$

The Greatest Common factor = ___

$\dfrac{(8 \div)}{(48 \div)} = \dfrac{}{}$

The Greatest Common factor = ___

$\dfrac{(8 \div)}{(32 \div)} = \dfrac{}{}$

The Greatest Common factor = ___

2. <u>Compare</u> fractions using ">", "<", or "=".

$\dfrac{3}{8}$ ... $\dfrac{5}{8}$ $\dfrac{4}{12}$ ... $\dfrac{9}{12}$

$\dfrac{5}{7}$ ... $\dfrac{2}{7}$ $\dfrac{3}{9}$ ... $\dfrac{7}{9}$

1. <u>Reduce</u> each fraction to its lowest terms and <u>color</u> the fraction of each shape. <u>Write</u> the missing numbers.

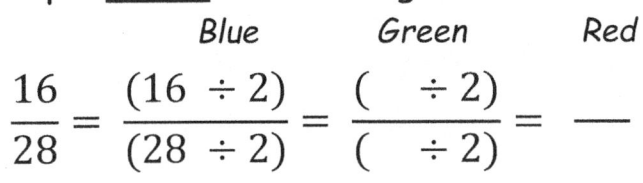

              Blue   Green   Red

$\dfrac{16}{28} = \dfrac{(16 \div 2)}{(28 \div 2)} = \dfrac{(\div 2)}{(\div 2)} = \dfrac{}{}$

The Greatest common factor is the greatest factor that divides both the numerator and the denominator evenly into a number, and there is no remainder.

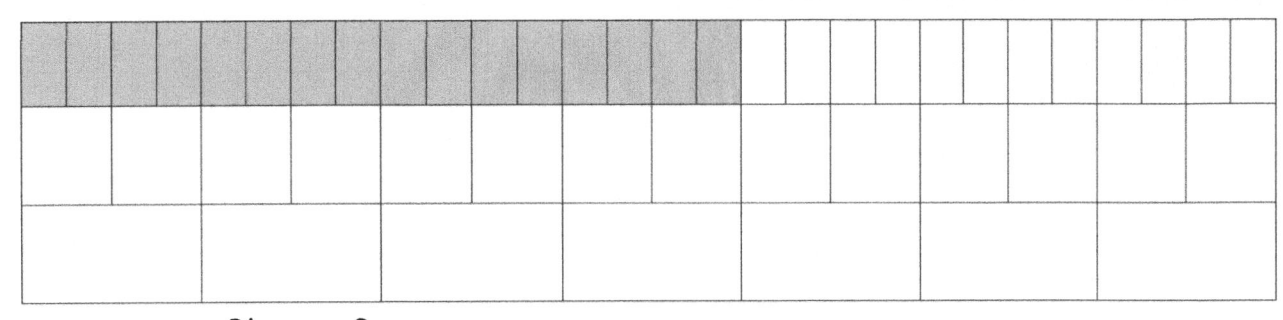

          Blue   Green

$\dfrac{20}{35} = \dfrac{(\div 5)}{(\div 5)} = \dfrac{}{}$

The common factor and the greatest common factor is 5.

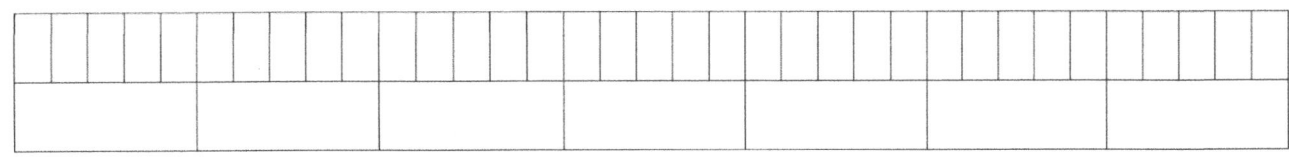

          Blue   Green

$\dfrac{14}{21} = \dfrac{(\div 7)}{(\div 7)} = \dfrac{}{}$

The common factor and the greatest common factor is 7.

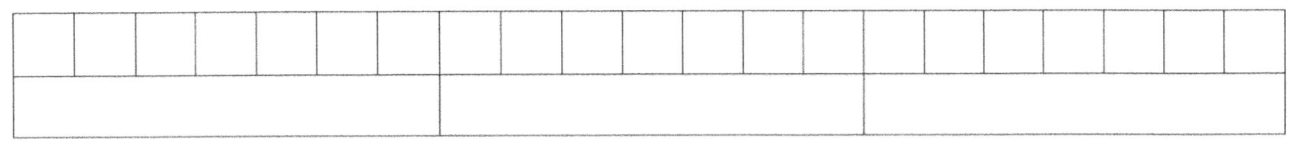

          Blue   Green   Red

$\dfrac{36}{40} = \dfrac{(\div 2)}{(\div 2)} = \dfrac{(\div 2)}{(\div 2)} = \dfrac{}{}$

Factors of 36: 1, 2, 3, 4, 6, 9, 12, 18, 36.
Factors of 40: 1, 2, 4, 5, 8, 10, 20, 40.
The **greatest common factor (GCF)** is 4.

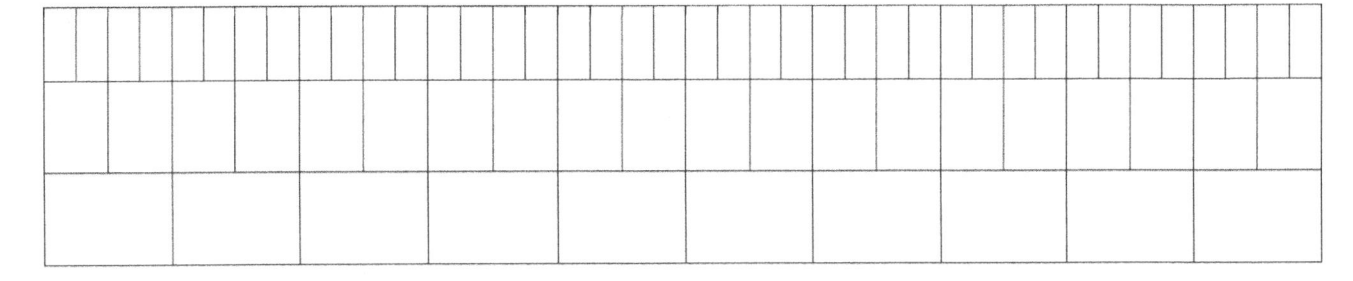

1. <u>Reduce</u> each fraction to its simplest form and <u>color</u> the fraction of each shape. <u>Write</u> the missing numbers.

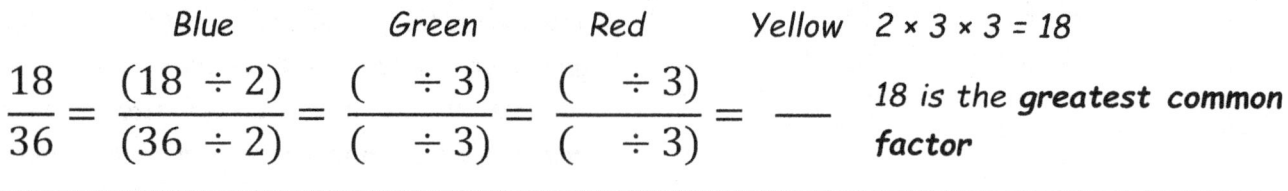

          Blue   Green   Red   Yellow 2 × 3 × 3 = 18

$$\frac{18}{36} = \frac{(18 \div 2)}{(36 \div 2)} = \frac{(\div 3)}{(\div 3)} = \frac{(\div 3)}{(\div 3)} = \frac{}{}$$

18 is the **greatest common factor**

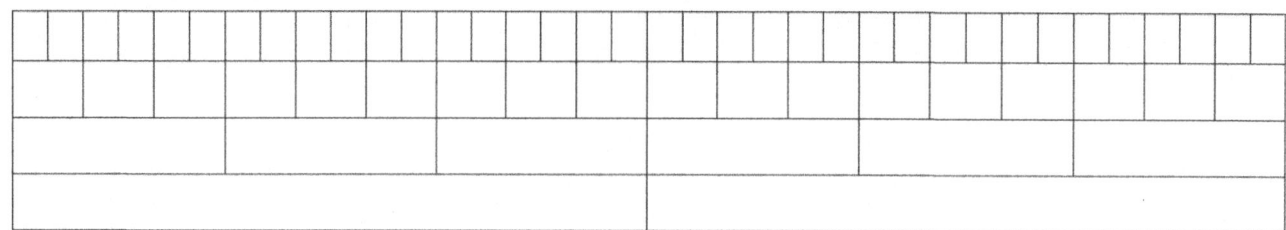

          Blue   Green   Red   Yellow 2 × 2 × 2 = __

$$\frac{32}{40} = \frac{(32 \div 2)}{(40 \div 2)} = \frac{(\div 2)}{(\div 2)} = \frac{(\div 2)}{(\div 2)} = \frac{}{}$$

__ is the **greatest common factor**

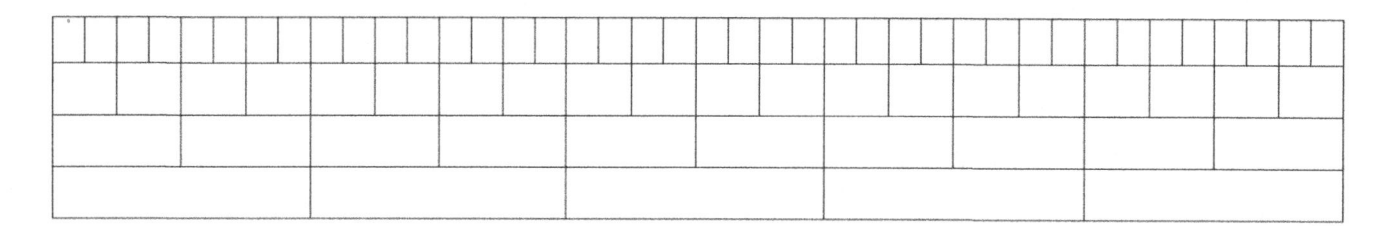

          Blue   Green   Red

$$\frac{28}{36} = \frac{(28 \div 2)}{(36 \div 2)} = \frac{(\div 2)}{(\div 2)} = \frac{}{}$$

2 × 2 = __

__ is the **greatest common factor**

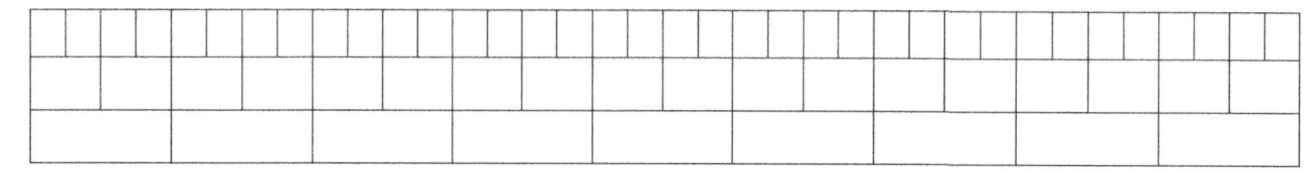

          Blue   Yellow

$$\frac{15}{18} = \frac{(15 \div)}{(18 \div)} = \frac{}{}$$

__ is the **greatest common factor**

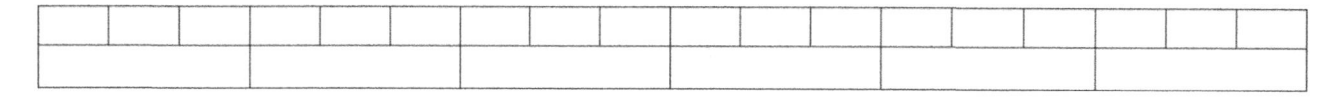

1. <u>Reduce</u> each fraction to its simplest form. <u>Write</u> the missing numbers.

$$\frac{32}{48} = \frac{(32 \div 2)}{(48 \div 2)} = \frac{(\div 2)}{(\div 2)} = \frac{(\div 2)}{(\div 2)} = \frac{(\div 2)}{(\div 2)} = \underline{}$$

$$\frac{(32 \div)}{(48 \div)} = \underline{} \qquad \underline{} \times \underline{} \times \underline{} \times \underline{} = \underline{} \text{ is the \textbf{\textit{greatest common factor}}}$$

$$\frac{12}{28} = \frac{(\div 2)}{(\div 2)} = \frac{(\div 2)}{(\div 2)} = \underline{} \qquad\qquad \frac{(12 \div)}{(28 \div)} = \underline{}$$

_____ is the *greatest common factor*

$$\frac{36}{48} = \frac{(\div 2)}{(\div 2)} = \frac{(\div 2)}{(\div 2)} = \frac{(\div 3)}{(\div 3)} = \underline{} \qquad \frac{(36 \div)}{(48 \div)} = \underline{}$$

_____ is the *greatest common factor*

$$\frac{50}{75} = \frac{(\div 5)}{(\div 5)} = \frac{(\div 5)}{(\div 5)} = \underline{} \qquad\qquad \frac{(50 \div)}{(75 \div)} = \underline{}$$

_____ is the *greatest common factor*

2. <u>Write</u> the fraction name under each picture. <u>Compare</u> the fractions.

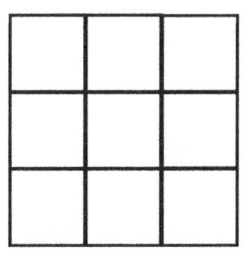

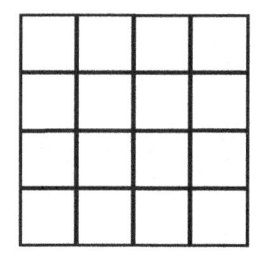

$$\frac{white\ blocks}{blocks\ in\ total} \qquad\qquad \frac{\cdots}{\cdots} \qquad\qquad \frac{\cdots}{\cdots}$$

1. <u>Reduce</u> each fraction to its simplest form. <u>Write</u> the missing numbers.

$$\frac{56}{72} = \frac{(\quad \div 2)}{(\quad \div \quad)} = \frac{(\quad \div \quad)}{(\quad \div 2)} = \frac{(\quad \div 2)}{(\quad \div \quad)} = \frac{\quad}{\quad} \qquad \frac{(56 \div \quad)}{(72 \div \quad)} = \frac{\quad}{\quad}$$

_____ is the **greatest common factor**

$$\frac{36}{81} = \frac{(\quad \div 3)}{(\quad \div \quad)} = \frac{(\quad \div 3)}{(\quad \div \quad)} = \frac{\quad}{\quad} \qquad \frac{(36 \div \quad)}{(81 \div \quad)} = \frac{\quad}{\quad}$$

_____ is the **greatest common factor**

$$\frac{60}{96} = \frac{(\quad \div 2)}{(\quad \div \quad)} = \frac{(\quad \div 2)}{(\quad \div \quad)} = \frac{(\quad \div 3)}{(\quad \div \quad)} = \frac{\quad}{\quad} \qquad \frac{(60 \div \quad)}{(96 \div \quad)} = \frac{\quad}{\quad}$$

_____ is the **greatest common factor**

$$\frac{48}{96} = \frac{(\quad \div 2)}{(\quad \div \quad)} = \frac{(\quad \div 2)}{(\quad \div \quad)} = \frac{(\quad \div 2)}{(\quad \div \quad)} = \frac{(\quad \div 2)}{(\quad \div \quad)} = \frac{(\quad \div 3)}{(\quad \div \quad)} = \frac{\quad}{\quad}$$

$$\frac{(48 \div \quad)}{(96 \div \quad)} = \frac{\quad}{\quad} \qquad \qquad \text{_____ is the \textbf{greatest common factor}}$$

2. <u>Write</u> the fraction name under each picture.
<u>Compare</u> fractions.

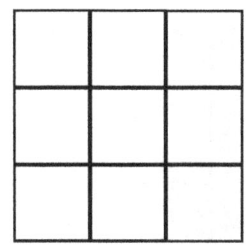

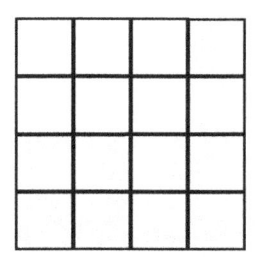

___   ___

1. <u>Reduce</u> each fraction to its simplest form. <u>Find</u> the greatest common factor (GCF).

$\frac{(20 \div)}{(28 \div)} = \frac{}{}$ $\frac{(18 \div)}{(27 \div)} = \frac{}{}$ $\frac{(12 \div)}{(42 \div)} = \frac{}{}$

GCF = _____   GCF = _____   GCF = _____

$\frac{(6 \div)}{(63 \div)} = \frac{}{}$ $\frac{(15 \div)}{(35 \div)} = \frac{}{}$ $\frac{(49 \div)}{(77 \div)} = \frac{}{}$

GCF = _____   GCF = _____   GCF = _____

$\frac{(24 \div)}{(56 \div)} = \frac{}{}$ $\frac{(16 \div)}{(36 \div)} = \frac{}{}$ $\frac{(20 \div)}{(32 \div)} = \frac{}{}$

GCF = _____   GCF = _____   GCF = _____

$\frac{(10 \div)}{(34 \div)} = \frac{}{}$ $\frac{(21 \div)}{(60 \div)} = \frac{}{}$ $\frac{(9 \div)}{(24 \div)} = \frac{}{}$

GCF = _____   GCF = _____   GCF = _____

$\frac{(25 \div)}{(45 \div)} = \frac{}{}$ $\frac{(40 \div)}{(64 \div)} = \frac{}{}$ $\frac{(35 \div)}{(84 \div)} = \frac{}{}$

GCF = _____   GCF = _____   GCF = _____

2. <u>Color</u> the fraction of each shape. <u>Compare</u> fractions.

$\frac{8}{9}$ ___ $\frac{8}{12}$

1. <u>Write in</u> the missing numbers and <u>cross out</u> the common prime factors.

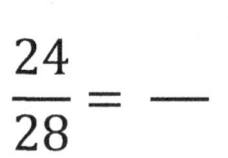

A prime number is a counting number that has exactly two different factors

$$\frac{24}{28} = \frac{}{}$$

Prime factors for 24:

2̶, 2̶, 2, ___

Prime factors for 28:

2̶, 2̶, ___

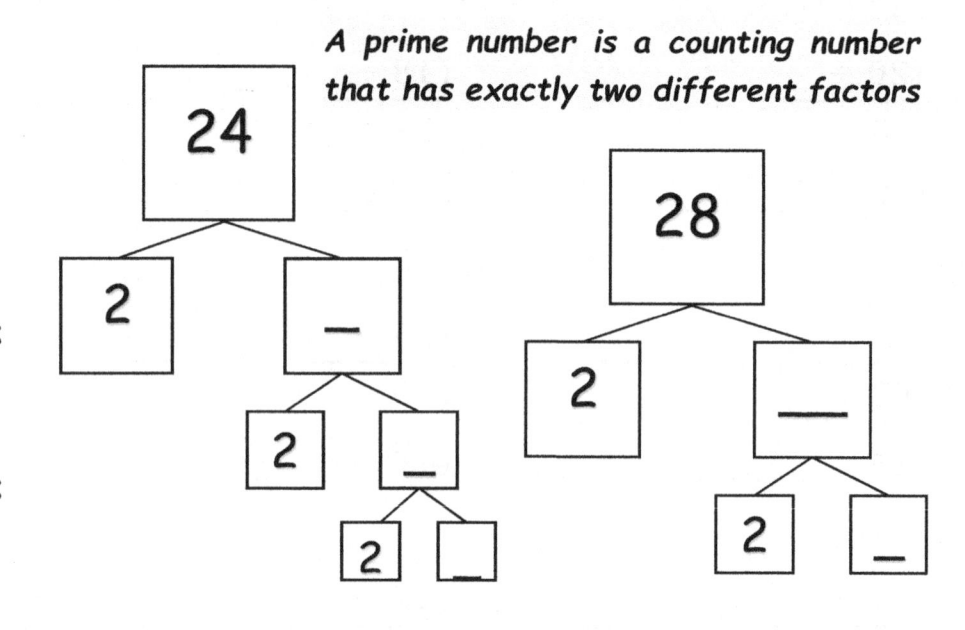

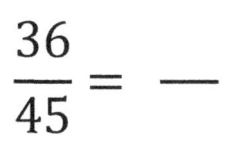

$$\frac{32}{40} = \frac{}{}$$

Prime factors for 32:

2, ___, ___, ___, ___

Prime factors for 40:

2, ___, ___, ___

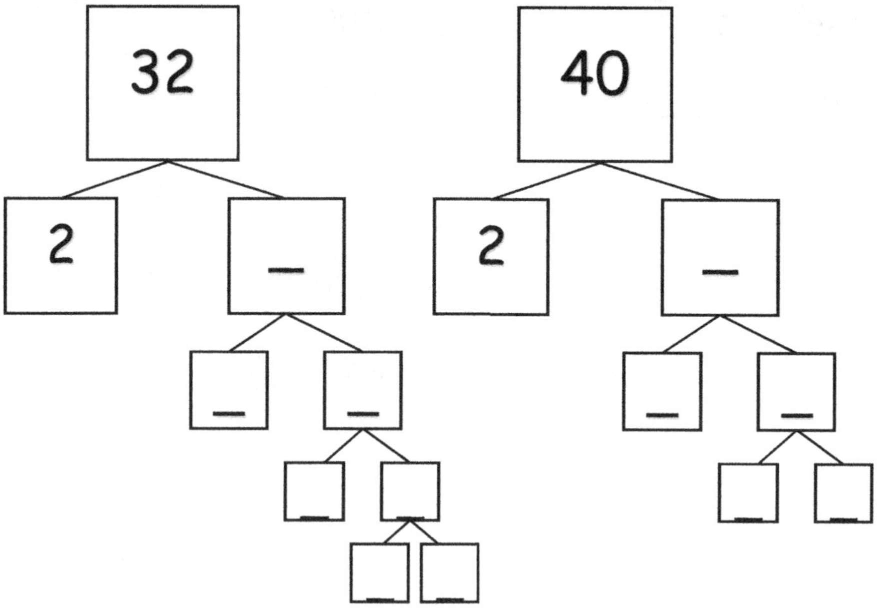

$$\frac{36}{45} = \frac{}{}$$

Prime factors for 36:

___, ___, ___, ___

Prime factors for 45:

5, ___, ___

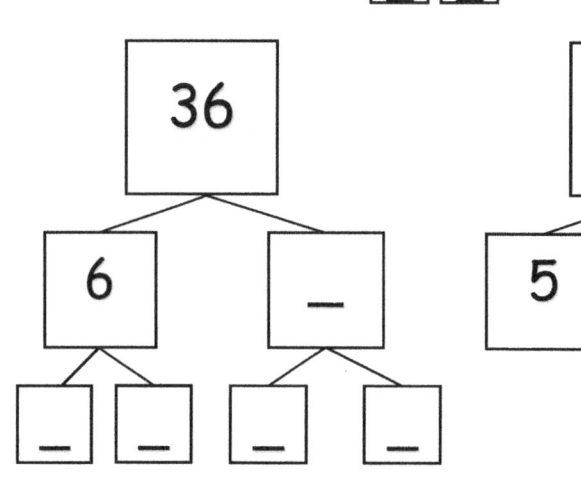

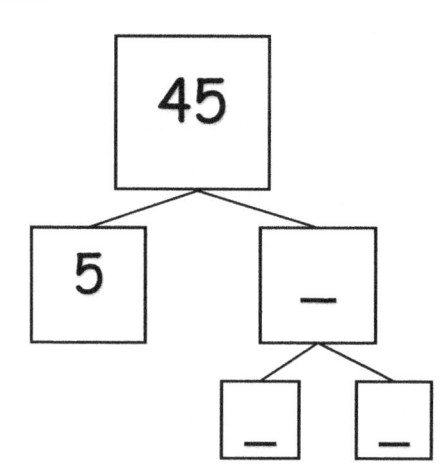

I have another idea! Grumpy, do you like weapons?

Yeah, I love Minecraft and Fortnite!

Great! Imagine, you're fighting with... Greeny, where are you?

Fighting? With Grumpy? Are you mad?

Maybe. Just imagine: you have a sword, a bow, and an axe.

$$\underline{\text{a sword, a bow, an axe}} \longleftarrow \text{Greeny}$$

Hint: *To simplify* means to **find both the numerator and the denominator that have no common prime factors** (or weapons).

I love this idea! So, I have a sword, a bow, an axe, and a trident.

$$\overline{\text{a sword, a bow, an axe, and a trident}} \longleftarrow \text{Grumpy}$$

To simplify means to **divide both the numerator and the denominator by the same number** (or weapon).

Fine! Now let's compare your weapons and cross out the weapons you BOTH have:

~~A sword~~, ~~a bow~~, and ~~an axe~~.

~~A sword~~, ~~a bow~~, ~~an axe~~, and a trident.

Hooray! I have a trident! So, guys, cross out the same weapon as your opponent has since you need only a unique weapon to win!

What a disgusting turn of events! Hate weapons! Love fractions! So, I need to find the factors in the numerator and the denominator and then, cross them out.

Factors? Ah, factors are the numbers, when multiplied together, they form a new number. We call that number a product, right?

Aha. For example, what are factors of 15? 15 = 5 × 3. So, 5 and 3 are factors of 15. Now, let's reduce your weapons' fraction to its lowest terms:

$$\frac{\text{a sword, a bow, an axe}}{\text{a sword, a bow, an axe, and a trident}} = \frac{\cancel{\text{a sword}}, \text{a bow, an axe}}{\cancel{\text{a sword}}, \text{a bow, an axe, and a trident}} =$$

$$= \frac{\cancel{\text{a bow}}, \text{an axe}}{\cancel{\text{a bow}}, \text{an axe, and a trident}} = \frac{\cancel{\text{an axe}}}{\cancel{\text{an axe}}, \text{and a trident}} = \frac{1}{\text{a trident}}$$

Can I try?

Hint: A prime number is any number that has only two factors, itself and 1. By the way, the number 1 is neither prime nor composite.

$$\frac{15}{35} = \frac{\cancel{5} \times 3}{\cancel{5} \times 7} = \frac{3}{7} \qquad \frac{16}{24} = \frac{\cancel{2} \times \cancel{2} \times \cancel{2} \times 2}{\cancel{2} \times \cancel{2} \times \cancel{2} \times 3} = \frac{2}{3}$$

Cancel the 5s in both the numerator and the denominator. *Cancel the 2s in both the numerator and the denominator. So, GCF is 8.*

1. <u>Reduce</u> each fraction to its simplest form. <u>Cross out</u> the same factors in the numerator and the denominator.

$$\frac{18}{28} = \frac{\cancel{2} \times 3 \times 3}{\cancel{2} \times 2 \times 7} = \frac{9}{14}$$ ← is in simplest form since 9 and 14 have only 1 as a common factor!

$$\frac{20}{30} = \frac{2 \times 2 \times 5}{2 \times 3 \times 5} = \frac{}{}$$

$$\frac{18}{22} = \frac{2 \times 3 \times 3}{2 \times 11} = \frac{}{}$$

$$\frac{16}{32} = \frac{2 \times 2 \times 2 \times 2}{2 \times 2 \times 2 \times 2 \times 2} = \frac{}{}$$

$$\frac{25}{40} = \frac{5 \times 5}{2 \times 2 \times 2 \times 5} = \frac{}{}$$

$$\frac{24}{48} = \frac{2 \times 2 \times 2 \times 3}{2 \times 2 \times 2 \times 2 \times 3} = \frac{}{}$$

$$\frac{30}{46} = \frac{2 \times 3 \times 5}{2 \times 23} = \frac{}{}$$

$$\frac{32}{48} = \frac{2 \times 2 \times 2 \times 2 \times 2}{2 \times 2 \times 2 \times 2 \times 3} = \frac{}{}$$

$$\frac{48}{64} = \frac{2 \times 2 \times 2 \times 2 \times 3}{2 \times 2 \times 2 \times 2 \times 2 \times 2} = \frac{}{}$$

$$\frac{44}{55} = \frac{2 \times 2 \times 11}{5 \times 11} = \frac{}{}$$

A fraction is written in its simplest form (= lowest term) when both the numerator and the denominator have NO common prime factors!

2. <u>Circle</u> the missing factor.

a) 10

b) 16

c) 14

d) 21

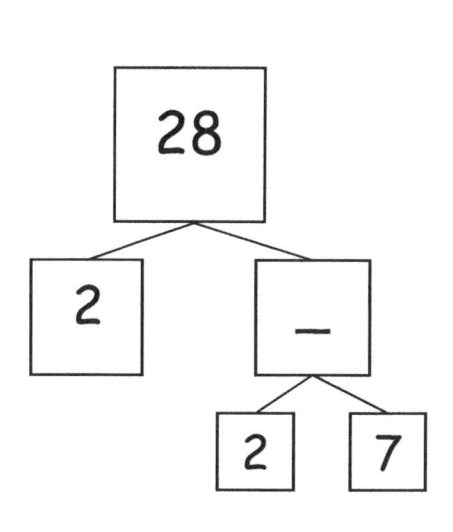

```
        [28]
        /  \
      [2]  [_]
           /  \
         [2]  [7]
```

1. <u>Write in</u> the missing numbers and <u>cross out</u> the common prime factors.

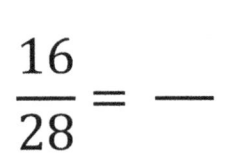

A composite number is a counting number that has more than two factors

$\dfrac{16}{28} = \dfrac{}{}$

Prime factors for 16:

2, ___, ___, ___

Prime factors for 28:

2, ___, ___

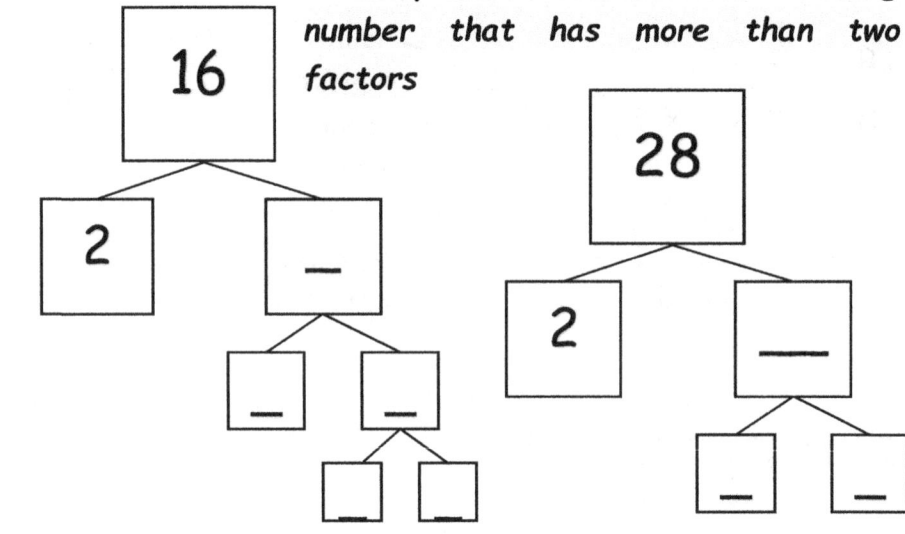

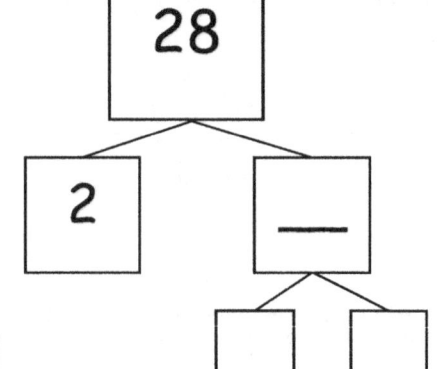

$\dfrac{48}{54} = \dfrac{}{}$

Prime factors for 48:

___, ___, ___, ___, ___

Prime factors for 54:

2, ___, ___, ___

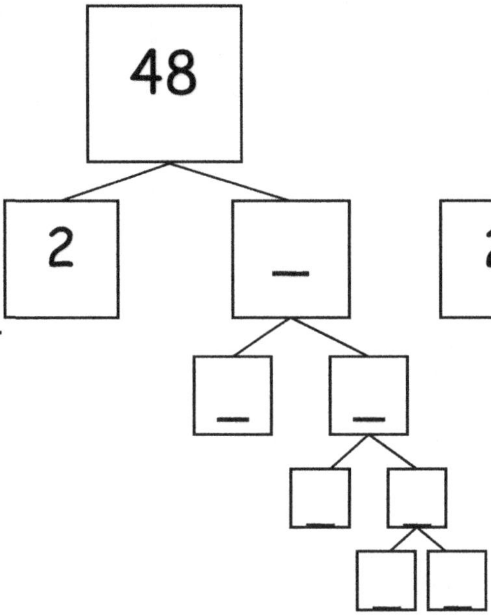

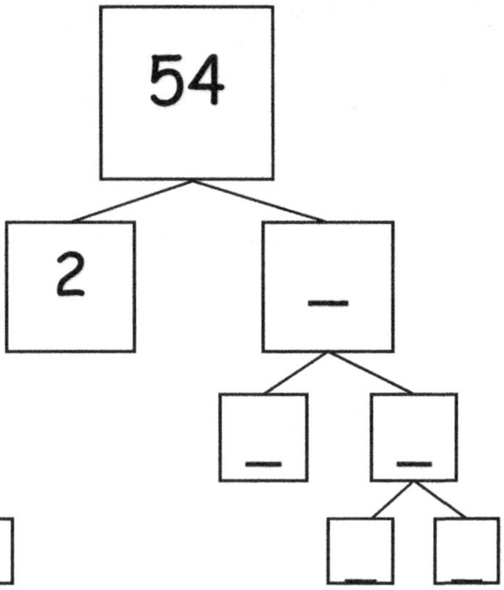

$\dfrac{16}{42} = \dfrac{}{}$

Prime factors for 16:

___, ___, ___, ___

Prime factors for 42:

2, ___, ___

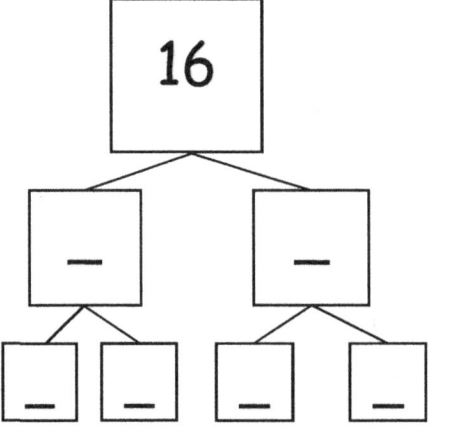

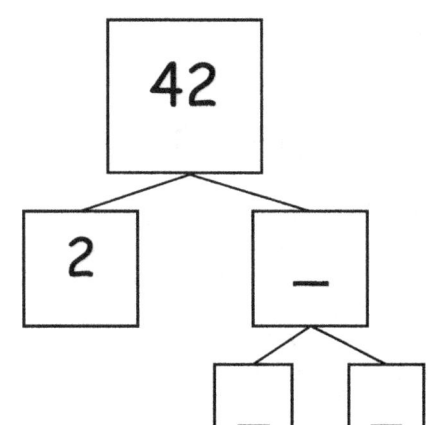

1. **Write in** the missing numbers.

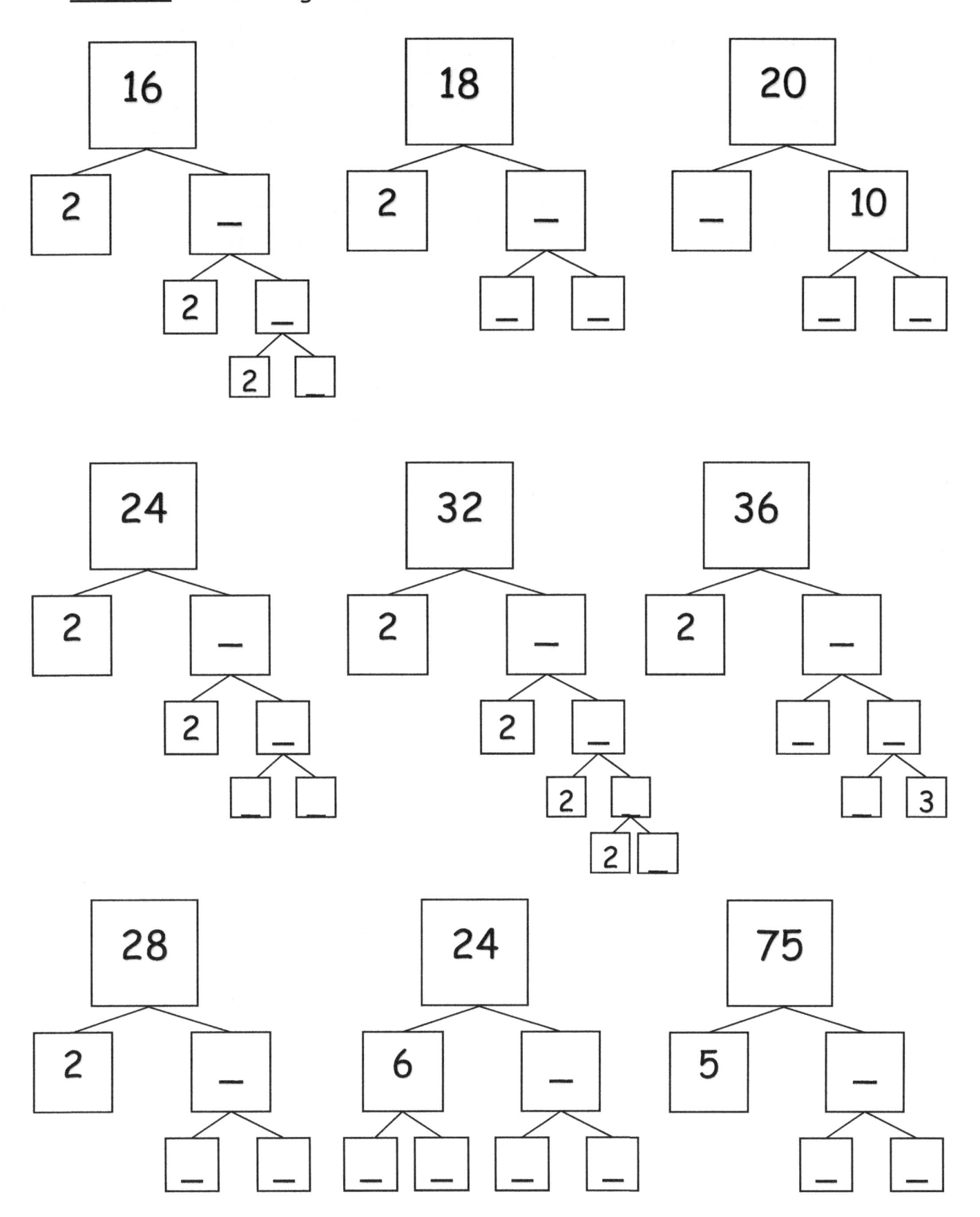

1. <u>Reduce</u> each fraction to simplest form. <u>Cross out</u> the same factors in the numerator and the denominator.

$\dfrac{28}{48} = \dfrac{\times \quad \times}{\times \quad \times \quad \times \quad \times} = \dfrac{}{}$

$\dfrac{36}{54} = \dfrac{\times \quad \times \quad \times}{\times \quad \times \quad \times} = \dfrac{}{}$

$\dfrac{52}{64} = \dfrac{\times \quad \times}{\times \quad \times \quad \times \quad \times \quad \times} = \dfrac{}{}$

$\dfrac{68}{75} = \dfrac{\times \quad \times \quad \times}{\times \quad \times \quad \times} = \dfrac{}{}$

$\dfrac{50}{70} = \dfrac{\times \quad \times \quad \times}{\times \quad \times \quad \times} = \dfrac{}{}$

$\dfrac{72}{80} = \dfrac{\times \quad \times \quad \times \quad \times}{\times \quad \times \quad \times \quad \times} = \dfrac{}{}$

$\dfrac{76}{84} = \dfrac{\times \quad \times \quad \times}{\times \quad \times \quad \times} = \dfrac{}{}$

$\dfrac{48}{88} = \dfrac{\times \quad \times \quad \times \quad \times}{\times \quad \times \quad \times} = \dfrac{}{}$

$\dfrac{80}{96} = \dfrac{\times \quad \times \quad \times \quad \times}{\times \quad \times \quad \times \quad \times \quad \times} = \dfrac{}{}$

$\dfrac{84}{100} = \dfrac{\times \quad \times \quad \times}{\times \quad \times \quad \times} = \dfrac{}{}$

2. <u>Circle</u> the missing factor(s).

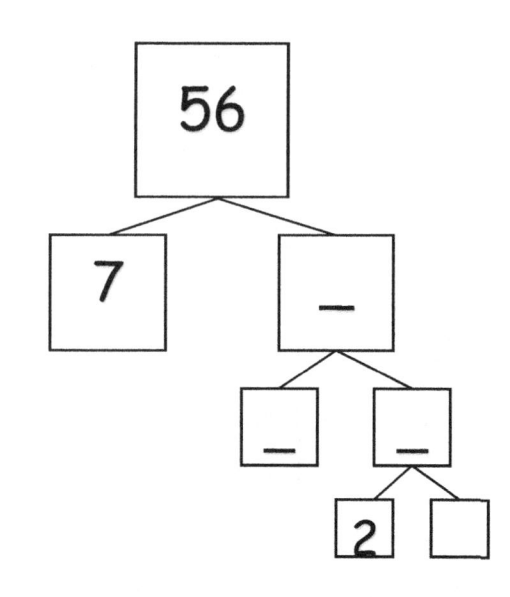

a) 2

b) 8

c) 6

d) 4

1. <u>Write in</u> the missing numbers.

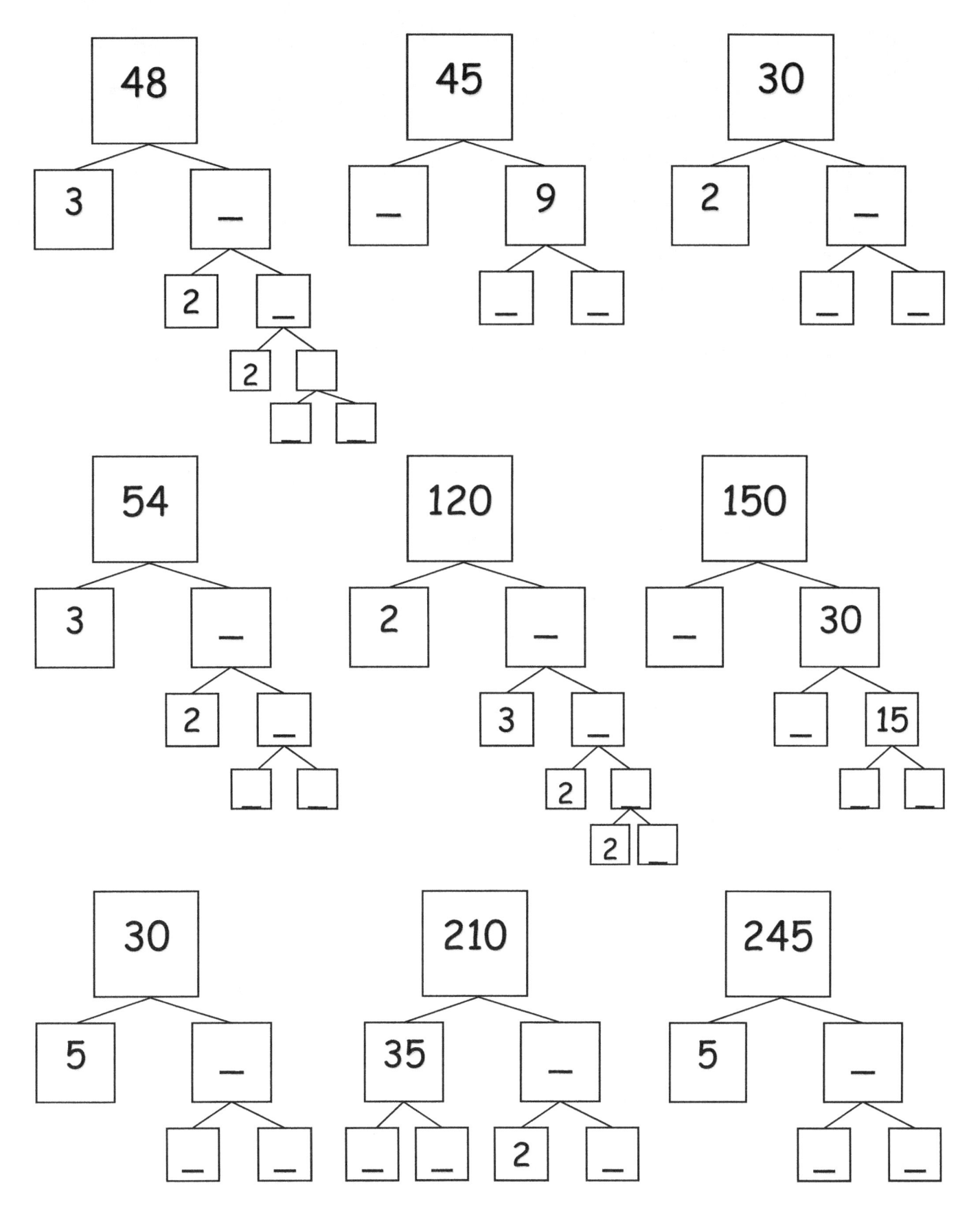

1. <u>Compare</u> fractions using "<," ">," or "=."

Help! SOS! Help!!!

Quick, Smarty, help him! I found a chest with diamonds! I want to play!

Congratulations, Grumpy! Purple-y, look, to compare fractions means to find out which fraction and smaller or larger.

When fractions have the same denominator, the fraction with the greatest (largest) numerator has the greatest value.

6 out of 16 squares are less than 9 out of 16 squares.

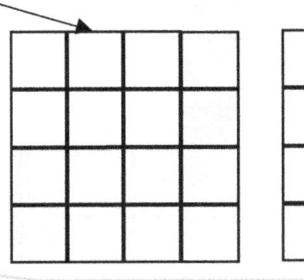

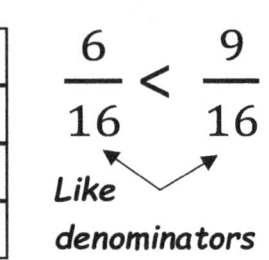

$$\frac{6}{16} < \frac{9}{16}$$

Like denominators

I compare the unlike fractions (= fractions with the different numerators or denominators). Let's start with fractions with like numerators but with unlike denominators:

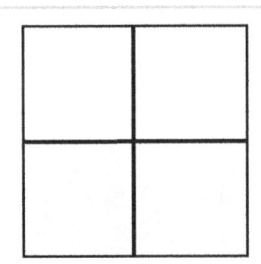

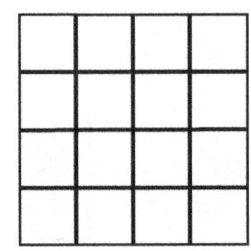

$$\frac{2}{4} \quad—\quad \frac{2}{16}$$

Unlike denominators

I divided my block into 4 huge equal parts and colored 2 parts white. Then, I divided another block into 16 tiny equal parts and colored 2 equal parts white. They look different!

Hm... If the fractions have like numerators, then, the fraction divided into a smaller number of equal parts is bigger.

1 part out of 2 equal parts is bigger than

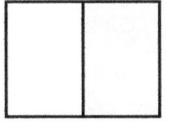

1 part out of 4 equal parts.

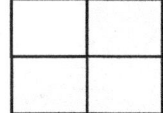

Look, I divided a candy into 4 equal parts and I ate 1 part out of the four equal parts. Or I divided a candy into 2 equal parts and I ate 1 part out of the two equal parts.

When did I eat more? Of course, when the candy was divided into 2 equal parts.

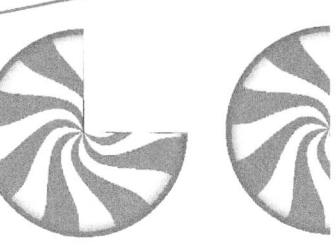

$$\frac{1}{4} < \frac{1}{2}$$

1 out of 4 equal parts is smaller than 1 out of 2 equal parts.

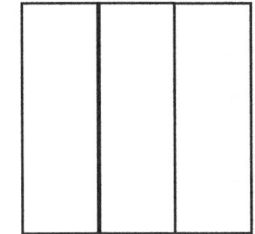

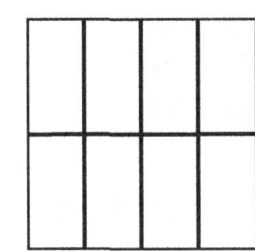

$$\frac{2}{3} > \frac{2}{8}$$

2 out of 3 equal parts are bigger than 2 out of 8 equal parts.

I divided my block into 3 huge equal parts and colored 2 parts white.

Then, I divided another block into 8 small equal parts and colored 2 equal parts white.

Which fraction is largest?

The larger fraction is the one with the smaller denominator (that means the whole is divided into a smaller number of equal parts).

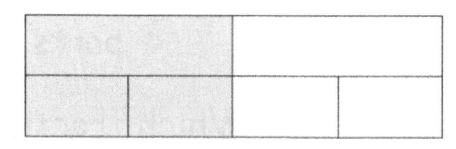

1. <u>Compare</u> fractions using "<," ">," or "=." <u>Color</u> the fraction of each shape. <u>Cross out</u> the wrong word. The first one is done for you.

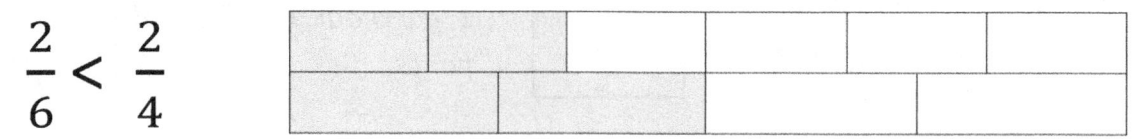

$\frac{2}{6} < \frac{2}{4}$

2 out of 6 equal parts are ~~bigger~~/smaller than 2 out of 4 equal parts.

$\frac{5}{8} \quad \frac{5}{11}$

5 out of 8 equal parts are bigger/smaller than 5 out of 11 equal parts.

$\frac{3}{5} \quad \frac{3}{3}$

3 out of 5 equal parts are bigger/smaller than 3 out of 3 equal parts.

$\frac{4}{8} \quad \frac{4}{12}$

4 out of 8 equal parts are bigger/smaller than 4 out of 12 equal parts.

2. <u>How many fourths</u> make a half? <u>How many sixths</u> make a half? <u>Write</u> the missing numbers. <u>Draw</u> and <u>color</u> the fraction of each shape.

$\frac{1}{2} = \frac{}{4}$

$\frac{1}{2} = \frac{}{}$

1. <u>Compare</u> fractions using "<," ">," or "=." <u>Color</u> the fraction of each shape. <u>Cross out</u> the wrong word.

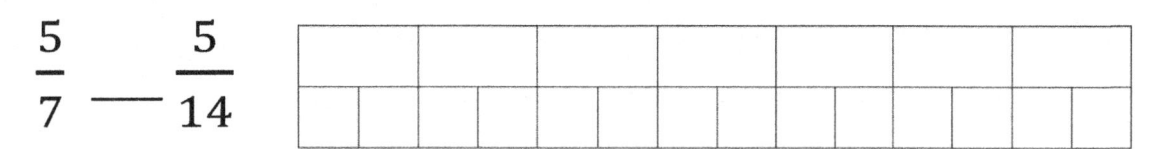

$\dfrac{5}{7}$ — $\dfrac{5}{14}$

5 out of 7 equal parts are bigger/smaller than 5 out of 14 equal parts.

$\dfrac{5}{16}$ — $\dfrac{5}{10}$

5 out of 16 equal parts are bigger/smaller than 5 out of 10 equal parts.

$\dfrac{6}{6}$ — $\dfrac{6}{10}$

6 out of 6 equal parts are bigger/smaller than 6 out of 10 equal parts.

$\dfrac{8}{20}$ — $\dfrac{8}{14}$

8 out of 20 equal parts are bigger/smaller than 8 out of 14 equal parts.

2. <u>How many tenths</u> make a half? <u>How many twelfths</u> make a quarter?

<u>Write</u> the missing numbers. <u>Draw</u> and <u>color</u> the fraction of each shape.

$\dfrac{1}{2} = \dfrac{}{}$

$\dfrac{1}{4} = \dfrac{}{}$

I'm a Minecraft fan. So, Steve meets $\frac{1}{3}$ of as many zombies as Alex. If Alex meets 12 zombies, how many zombies does Steve meet?

You need to find parts of a number. Remember that the word OF in fractions means MULTIPLY. So, Steve meets $\frac{1}{3}$ of 12 zombies in Math means $\frac{1}{3} \times 12$.

I have an idea! I need to divide my rectangle into 12 equal parts and color $\frac{1}{3}$ of them brown.

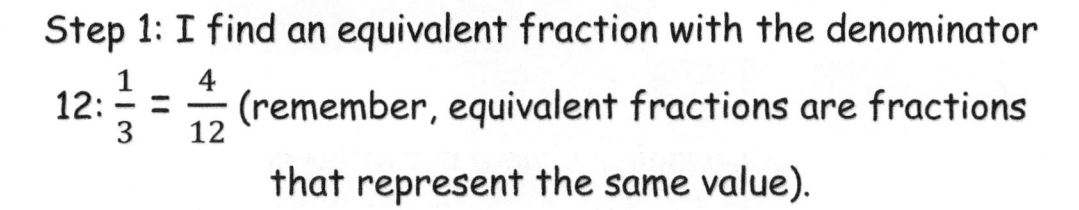

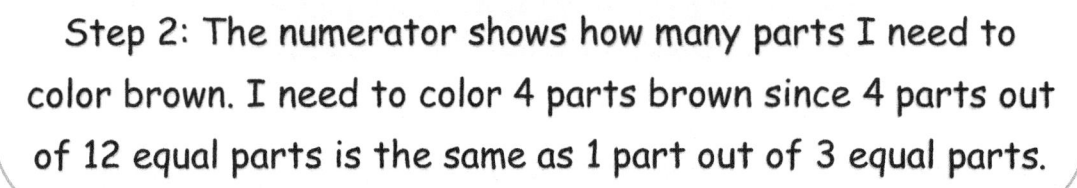

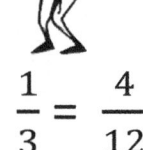

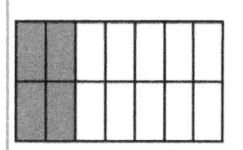

Step 1: I find an equivalent fraction with the denominator 12: $\frac{1}{3} = \frac{4}{12}$ (remember, equivalent fractions are fractions that represent the same value).

Step 2: The numerator shows how many parts I need to color brown. I need to color 4 parts brown since 4 parts out of 12 equal parts is the same as 1 part out of 3 equal parts.

$\frac{1}{3} = \frac{4}{12}$

Another algorithm:

Step 1: OF means multiply: $\quad \frac{1}{3} \times 12$

Step 2: write 12 as $\frac{12}{1}$: $\quad \frac{1}{3} \times \frac{12}{1}$

Step 3: Multiply numerators and denominators, cancel the 3s:

$$\frac{1}{3} \times 12 = \frac{1}{3} \times \frac{12}{1} = \frac{1 \times 12}{3 \times 1} = \frac{1 \times \cancel{12}^{4}}{\cancel{3} \times 1} = \frac{1 \times 4}{1 \times 1} = 4$$

Aha… I have another idea! All the zombies are divided equally into 3 equal groups. And I meet only $\frac{1}{3}$ of them.

The question is: If there are 12 zombies altogether, how many zombies are in one of the 3 equal parts?

Step 1: find how many zombies are in one of the three equal parts: $12 \div 3 = 4$

(4 zombies are in each of the three equal groups)

Step 2: find how many zombies are in one group: $1 \times 4 = 4$

(4 zombies in 1 group).

I wonder, if I meet $\boxed{\frac{2}{3}}$ of $\boxed{12}$ zombies, <u>how many zombies</u> do I meet?

That's the question I like. My favorite algorithm:

$$\frac{2}{3} \times 12 = \frac{2}{3} \times \frac{12}{1} = \frac{2 \times 12}{3 \times 1} = \frac{2 \times \cancel{12}}{\cancel{3} \times 1} = \frac{2 \times 4}{1 \times 1} = 8 \text{ (zombies)}$$

I like to solve it slower: $\frac{2}{3} \times 12$

Step 1: $12 \div 3 = 4$ (find $\frac{1}{3}$ out of 12)

(4 zombies in each of the three equal groups)

Step 2: $2 \times 4 = 8$ (find how many zombies are in 2 groups)

(8 zombies in 2 groups).

1. <u>Multiply</u> and <u>simplify</u> if possible. <u>Color</u> the correct number of images.

Find $\frac{1}{5}$ of $\boxed{15}$ flowers.

$\frac{1}{5} \times 15 = \frac{}{} \times \frac{}{} = \frac{ \times }{ \times } = \underline{}$

Find $\frac{3}{4}$ of $\boxed{16}$ stars.

Step 1: ___ ÷ ___ = ___ (find $\frac{1}{4}$ out of 16)

Step 2: ___ × ___ = ___

Find $\frac{5}{7}$ of $\boxed{21}$ flowers.

$\frac{}{} \times \underline{} = \frac{}{} \times \frac{}{} = \frac{ \times }{ \times } = \underline{}$

Find $\frac{5}{8}$ of $\boxed{24}$ stars.

Step 1: ___ ÷ ___ = ___ (find $\frac{1}{8}$ out of 24)

Step 2: ___ × ___ = ___

1. <u>Reduce</u> each fraction to simplest form, <u>write</u> the missing numbers, and <u>compare</u> fractions using "<," ">," or "=." <u>Color</u> the fraction of each shape. <u>Cross out</u> the wrong word.

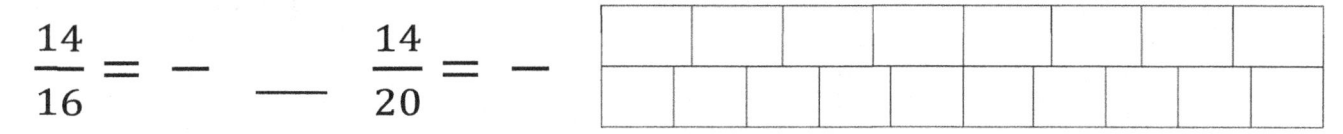

$\dfrac{14}{16} = \dfrac{\ }{\ }$ ___ $\dfrac{14}{20} = \dfrac{\ }{\ }$

__ out of __ equal parts are bigger/smaller than __ out of __ equal parts.

$\dfrac{16}{18} = \dfrac{\ }{\ }$ ___ $\dfrac{16}{22} = \dfrac{\ }{\ }$

__ out of __ equal parts are bigger/smaller than __ out of __ equal parts.

$\dfrac{20}{24} = \dfrac{\ }{\ }$ ___ $\dfrac{10}{14} = \dfrac{\ }{\ }$

__ out of __ equal parts are bigger/smaller than __ out of __ equal parts.

2. <u>Find and circle</u> the area of a rectangle. Hint: Figures are not drawn to scale.

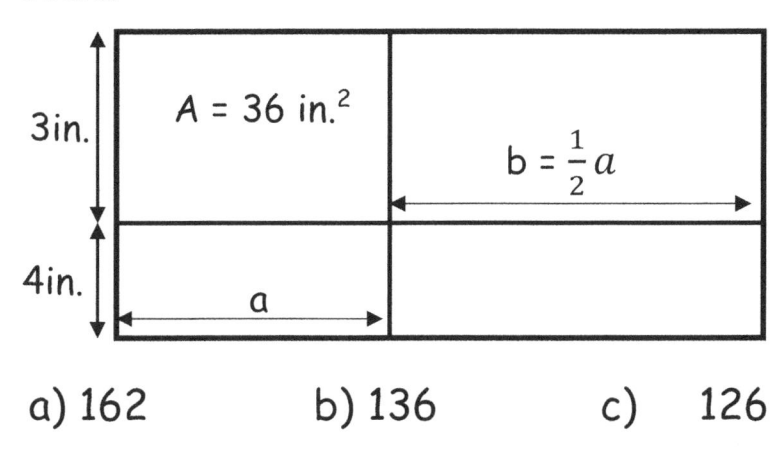

3in. | A = 36 in.² | b = $\dfrac{1}{2}a$
4in. | a

a) 162   b) 136   c)   126

1. <u>Write in</u> the missing numbers and <u>cross out</u> the common prime factors.

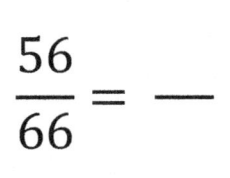

A prime number is a counting number that has exactly two different factors

$\dfrac{56}{66} = \dfrac{}{}$

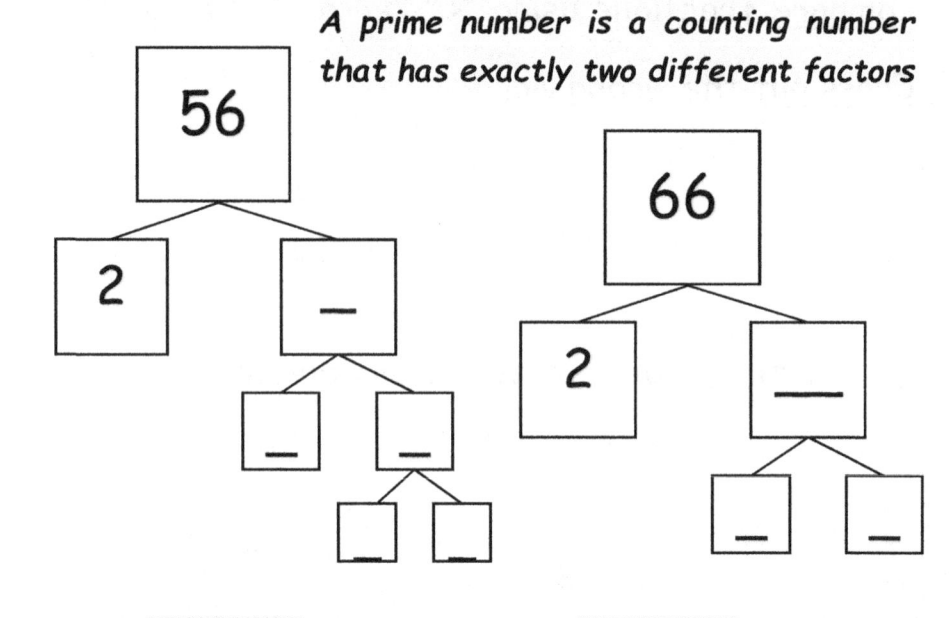

Prime factors for 56:

2, ___, ___, ___,

Prime factors for 66:

2, ___, ___

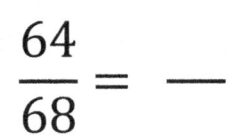

$\dfrac{64}{68} = \dfrac{}{}$

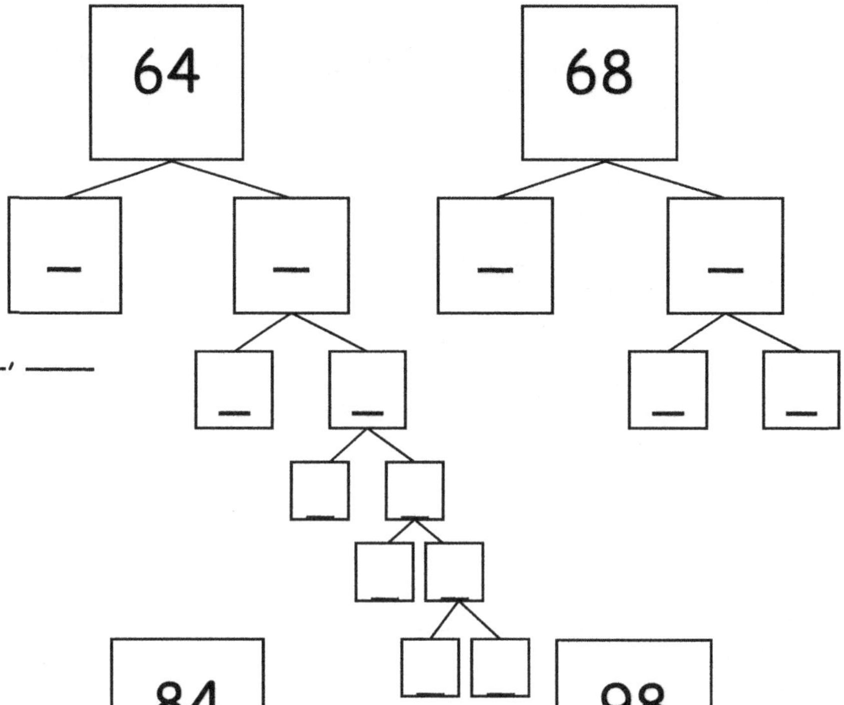

Prime factors for 64:

___, ___, ___, ___, ___, ___

Prime factors for 68:

___, ___, ___

$\dfrac{84}{98} = \dfrac{}{}$

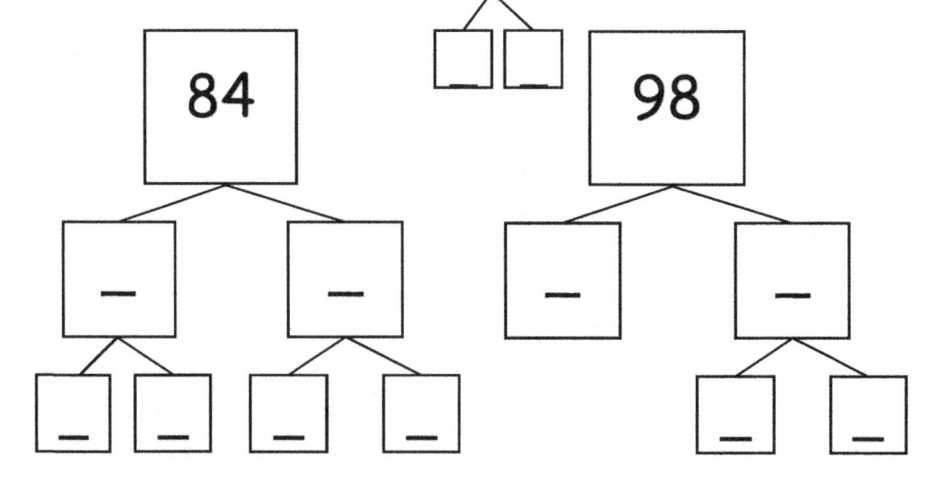

Prime factors for 84:

___, ___, ___, ___

Prime factors for 98:

___, ___, ___

1. <u>Reduce</u> each fraction to simplest form, <u>write</u> the missing numbers, and <u>compare</u> fractions using "<," ">," or "=." <u>Color</u> the fraction of each shape. <u>Cross out</u> the wrong word.

$\frac{5}{20} = \frac{}{}$ ___ $\frac{5}{10} = \frac{}{}$

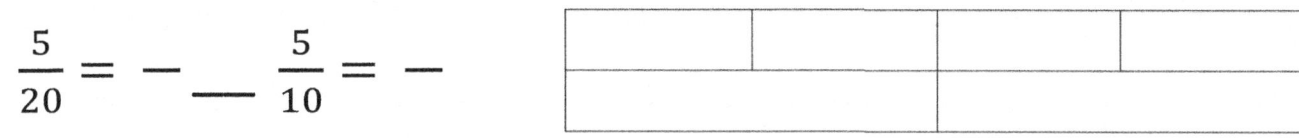

___ out of ___ equal parts are bigger/smaller than ___ out of ___ equal parts.

$\frac{6}{8} = \frac{}{}$ ___ $\frac{6}{10} = \frac{}{}$

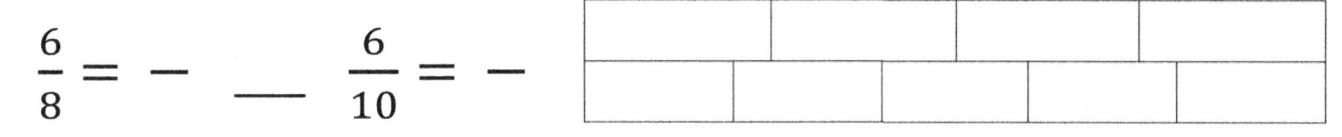

___ out of ___ equal parts are bigger/smaller than ___ out of ___ equal parts.

$\frac{4}{12} = \frac{}{}$ ___ $\frac{4}{8} = \frac{}{}$

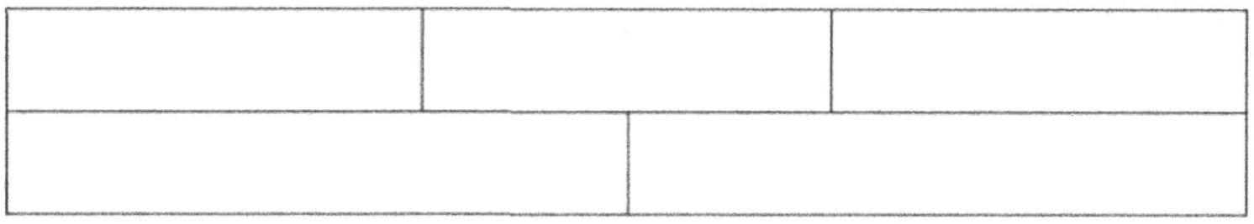

___ out of ___ equal parts are bigger/smaller than ___ out of ___ equal parts.

$\frac{2}{8} = \frac{}{}$ ___ $\frac{2}{12} = \frac{}{}$

___ out of ___ equal parts are bigger/smaller than ___ out of ___ equal parts.

2. <u>Write</u> 4 fractions that you cannot reduce to their simplest forms.

$\frac{}{}$; $\frac{}{}$; $\frac{}{}$; $\frac{}{}$; $\frac{}{}$

1. <u>Write in</u> the missing numbers.

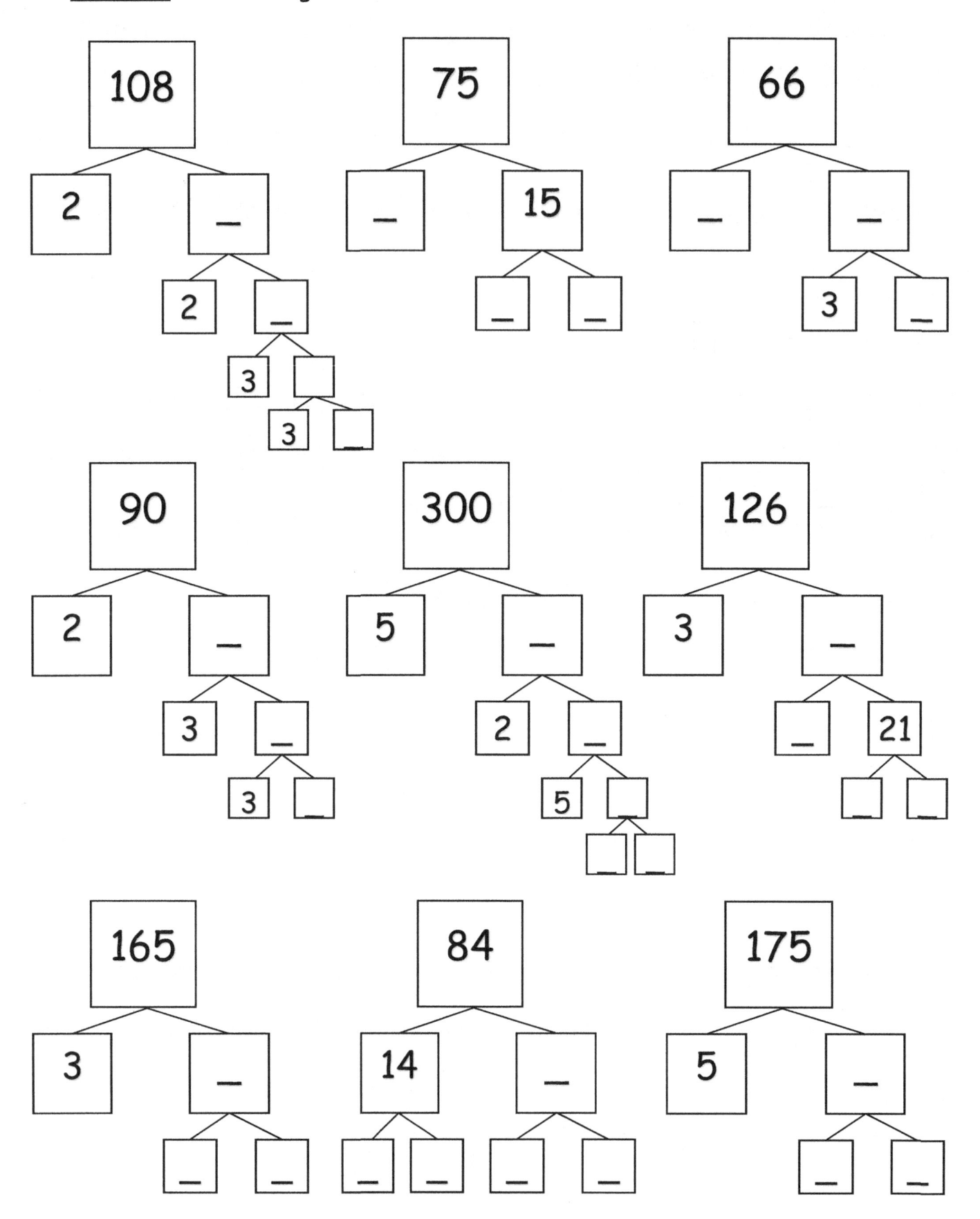

1. <u>Write</u> equivalent fractions. <u>Fill in</u> the missing numbers.

$\dfrac{2 \times}{3 \times} = \dfrac{}{6}$ $\dfrac{1 \times}{2 \times} = \dfrac{}{8}$

Common factor = 6 ÷ 3 = 2   Common factor = 8 ÷ 2 = 4

$\dfrac{3 \times}{4 \times} = \dfrac{}{12}$ $\dfrac{1 \times}{3 \times} = \dfrac{}{18}$

Common factor = _____   Common factor = _____

$\dfrac{7 \times}{8 \times} = \dfrac{}{24}$ $\dfrac{2 \times}{3 \times} = \dfrac{}{12}$

Common factor = _____   Common factor = _____

$\dfrac{5 \times}{6 \times} = \dfrac{}{36}$ $\dfrac{5 \times}{7 \times} = \dfrac{}{28}$

Common factor = _____   Common factor = _____

$\dfrac{4 \times}{5 \times} = \dfrac{}{35}$ $\dfrac{2 \times}{7 \times} = \dfrac{}{42}$

Common factor = _____   Common factor = _____

$\dfrac{3 \times}{5 \times} = \dfrac{}{60}$ $\dfrac{4 \times}{9 \times} = \dfrac{}{45}$

Common factor = _____   Common factor = _____

$\dfrac{2 \times}{5 \times} = \dfrac{}{30}$ $\dfrac{1 \times}{8 \times} = \dfrac{}{48}$

Common factor = _____   Common factor = _____

$\dfrac{2 \times}{3 \times} = \dfrac{}{60}$ $\dfrac{1 \times}{4 \times} = \dfrac{}{40}$

Common factor = _____   Common factor = _____

2. In a florist, $\dfrac{1}{6}$ of the flowers are daisies and the rest are roses. If there are 30 flowers in all, <u>find</u> the roses.   Answer: _____.

1. <u>Multiply</u> and <u>simplify</u> if possible.

<u>Find</u> and <u>color</u> $\frac{2}{3}$ of 21 blocks.

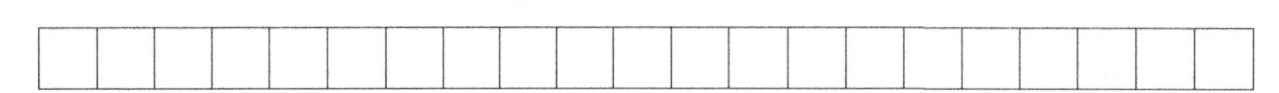

$\frac{2}{3} \times 21 = \frac{}{} \times \frac{}{} = \frac{ \times }{ \times } = \underline{}$

<u>Find</u> and <u>color</u> $\frac{6}{11}$ of 33 blocks.

Step 1: ____ ÷ ____ = ____ (find $\frac{1}{11}$ out of 33)

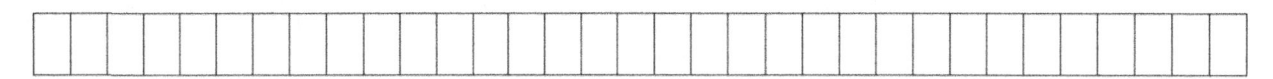

Step 2: ____ × ____ = ____

<u>Find</u> and <u>color</u> $\frac{7}{9}$ of 27 blocks.

$\frac{7}{9} \times 27 = \frac{}{} \times \frac{}{} = \frac{ \times }{ \times } = \underline{}$

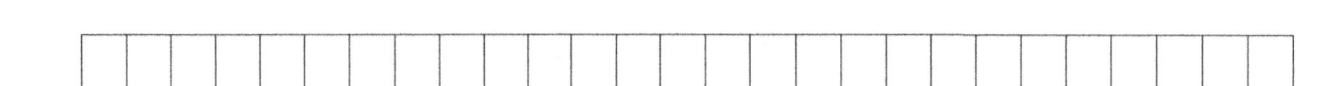

<u>Find</u> and <u>color</u> $\frac{4}{7}$ of 28 blocks.

Step 1: ____ ÷ ____ = ____ (find $\frac{1}{7}$ out of 28)

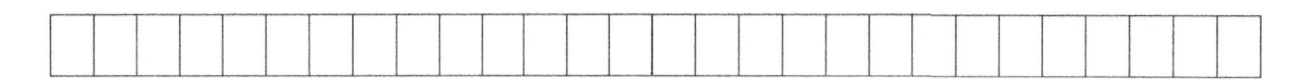

Step 2: ____ × ____ = ____

1. <u>Reduce</u> each fraction to simplest form, <u>write</u> the missing numbers, and <u>compare</u> fractions using "<," ">," or "=." <u>Color</u> the fraction of each shape. <u>Cross out</u> the wrong word.

$\frac{36}{48} = \frac{}{} \quad \underline{} \quad \frac{36}{60} = \frac{}{}$

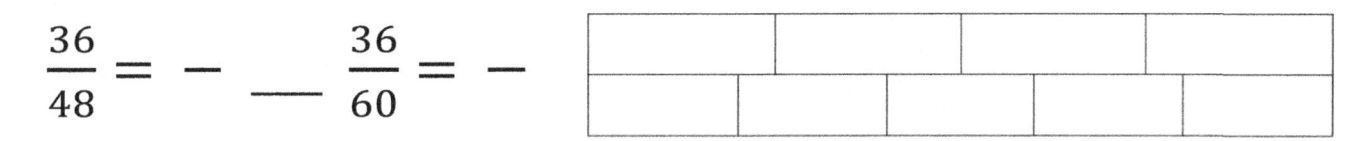

__ out of __ equal parts are bigger/smaller than __ out of __ equal parts.

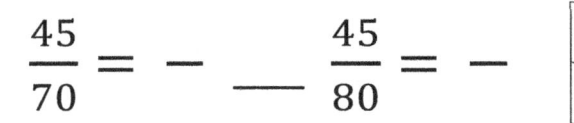

$\frac{45}{70} = \frac{}{} \quad \underline{} \quad \frac{45}{80} = \frac{}{}$

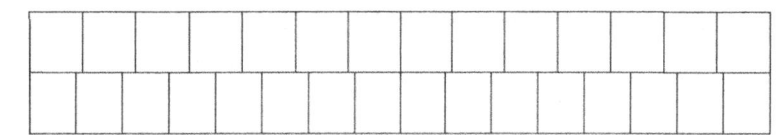

__ out of __ equal parts are bigger/smaller than __ out of __ equal parts.

$\frac{40}{64} = \frac{}{} \quad \underline{} \quad \frac{60}{108} = \frac{}{}$

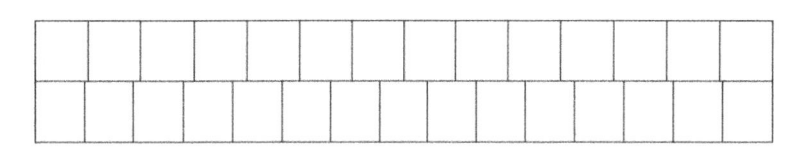

__ out of __ equal parts are bigger/smaller than __ out of __ equal parts.

$\frac{28}{49} = \frac{}{} \quad \underline{} \quad \frac{28}{63} = \frac{}{}$

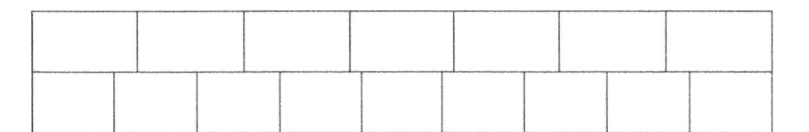

__ out of __ equal parts are bigger/smaller than __ out of __ equal parts.

2. <u>Find</u> d and <u>circle</u> the correct answer. Hint: Figures are not drawn to scale.

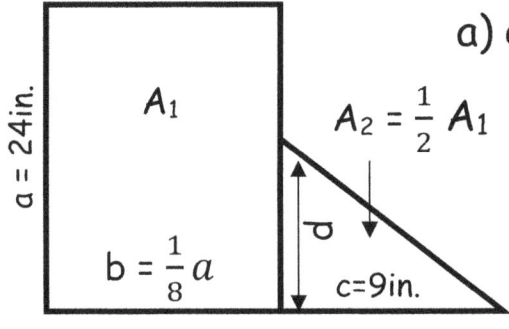

a) d = 9   b) d = 8   c) d = 7

$a = 24$ in.
A_1
$b = \frac{1}{8} a$
$A_2 = \frac{1}{2} A_1$
$c = 9$ in.

Math Success Practice Workbook Fractions 4th 5th 6th Grades

1. <u>Multiply</u> and <u>simplify</u> if possible. <u>Circle</u> the right answer (letter).

<u>Find</u> and <u>color</u> $\frac{6}{11}$ of 22 blocks.

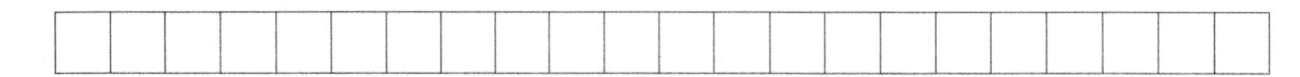

$\frac{6}{11} \times 22 = \frac{}{} \times \frac{}{} = \frac{ \times }{ \times } = \underline{}$

a) 10
b) 12
c) 15

<u>Find</u> and <u>color</u> $\frac{7}{10}$ of 30 blocks.

Step 1: ___ ÷ ___ = ___ (find $\frac{1}{10}$ out of 30)

Step 2: ___ × ___ = ___

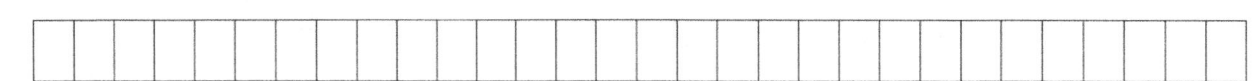

a) 21
b) 35
c) 20

<u>Find</u> and <u>color</u> $\frac{3}{8}$ of 32 blocks.

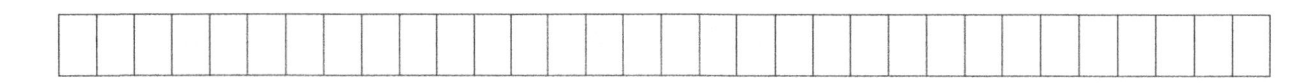

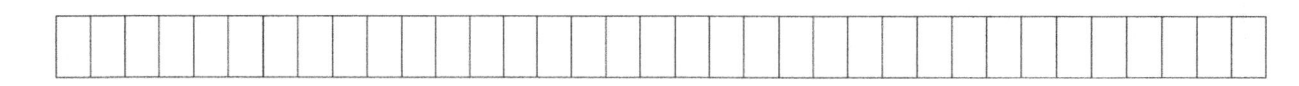

$\frac{3}{8} \times 32 = \frac{}{} \times \frac{}{} = \frac{ \times }{ \times } = \underline{}$

a) 16
b) 10
c) 12

<u>Find</u> and <u>color</u> $\frac{5}{7}$ of 35 blocks.

Step 1: ___ ÷ ___ = ___ (find $\frac{1}{7}$ out of 35)

Step 2: ___ × ___ = ___

a) 20
b) 35
c) 25

50   © 2020 Home Run Press, LLC

1. Reduce each fraction to simplest form, write the missing numbers, and compare fractions using "<," ">," or "=." Color the fraction of each shape. Cross out the wrong word.

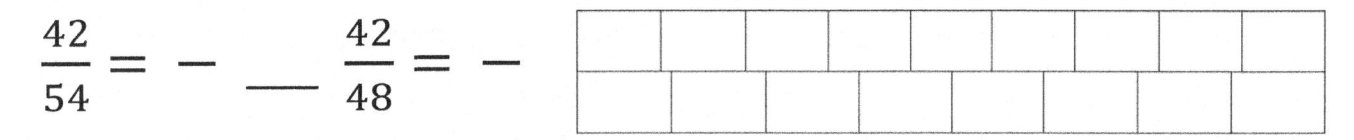

$\frac{42}{54} = \frac{\ }{\ }$ ___ $\frac{42}{48} = \frac{\ }{\ }$

__ out of __ equal parts are bigger/smaller than __ out of __ equal parts.

$\frac{24}{40} = \frac{\ }{\ }$ ___ $\frac{24}{64} = \frac{\ }{\ }$

__ out of __ equal parts are bigger/smaller than __ out of __ equal parts.

$\frac{20}{24} = \frac{\ }{\ }$ ___ $\frac{40}{64} = \frac{\ }{\ }$

__ out of __ equal parts are bigger/smaller than __ out of __ equal parts.

2. Find and circle the area of a rectangle. Hint: Figures are not drawn to scale.

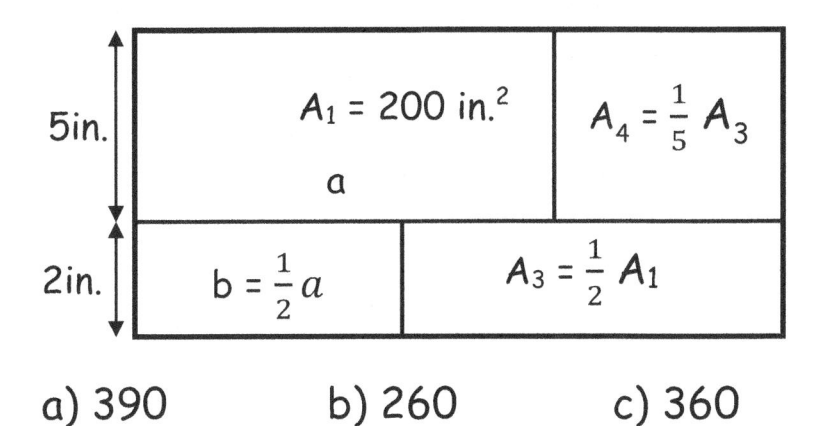

5in. $A_1 = 200$ in.^{2} $A_4 = \frac{1}{5} A_3$
       a
2in. $b = \frac{1}{2} a$ $A_3 = \frac{1}{2} A_1$

a) 390   b) 260   c) 360

1. <u>Compare</u> fractions with unlike numerators and denominators.

Well, I cannot compare fractions until they have been changed into equivalent fractions, right?

Yeah, you need fractions with the same denominators!

Okay, first, fractions with different denominators are called unlike fractions.

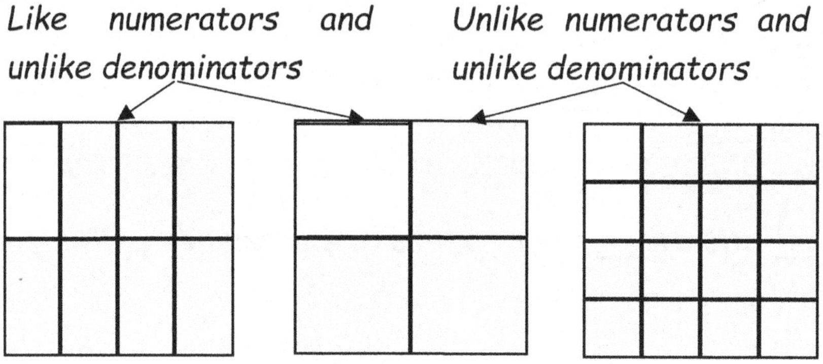

Like numerators and unlike denominators

Unlike numerators and unlike denominators

$\frac{1}{8}$ $\frac{1}{4}$ $\frac{2}{16}$

Hints: If a whole is divided into more equal parts (12 > 8), each equal part will be smaller ($\frac{1}{8} > \frac{1}{12}$.)

The larger the number shown by the denominator (5 > 2), the smaller is the size of this denominator ($\frac{1}{5} < \frac{1}{2}$.)

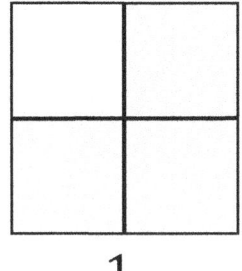

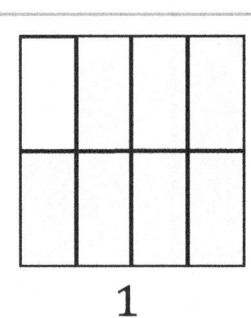

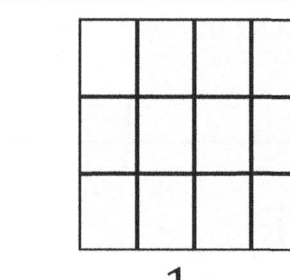

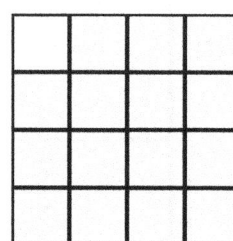

$\frac{1}{4}$ > $\frac{1}{8}$ > $\frac{1}{12}$ > $\frac{1}{16}$

The question is: How can I compare unlike fractions with unlike denominators and unlike numerators?

☺ Hint: Find **equivalent fractions with the same denominator**.

Look, I divided a candy into 4 equal parts and I ate 3 parts out of the four equal parts. Or I divided a candy into 2 equal parts and I ate 1 part out of the two equal parts.

When did I eat more?

Change the original denominators by multiplying both the numerator and the denominator by the same number to get equivalent fractions.

Which fraction is bigger?

$$\frac{3}{4} \; ? \; \frac{1}{2}$$

To compare unlike fractions – for example, $\frac{3}{4}$ and $\frac{5}{8}$, I need to change $\frac{3}{4}$ to an equivalent fraction with the denominator 8. Thus, I can compare the two fractions. To do that, I need to find an equivalent fraction to $\frac{3}{4}$:

$$\frac{3}{4} = \frac{6}{8}$$

Multiply both the numerator and the denominator by the common factor 2: 3 × 2 = 6; 4 × 2 = 8

$$\frac{6}{8} > \frac{5}{8}$$

These fractions are equivalent in size since they represent the whole divided into 8 equal parts

The same denominator for both fractions is called a **common denominator**. Now, I can easily compare the fractions:

$$\frac{3}{4} > \frac{5}{8}$$

Help me compare these unlike fractions:

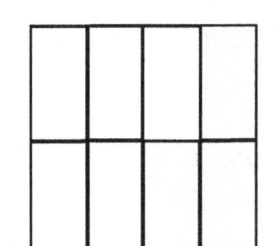

$$\frac{5}{8} \; ? \; \frac{6}{16}$$

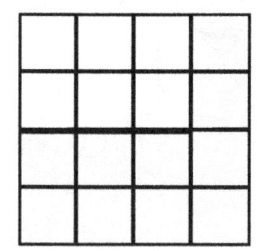

Step 1: Find the Least Common Denominator (LCD). Change one or both fractions into equivalent fractions.

Multiples of 8: 8, **16**, 24, **32**, 40, **48**, 56, **64**, ...

Multiples of 16: **16**, 32, 48, 64, 80, 96, ...

The **Least Common Denominator** is 16

$$\frac{5}{8} \quad \frac{7}{16} \longrightarrow \frac{5 \times 2}{8 \times 2} \quad \frac{7}{16} \longrightarrow \frac{10}{16} \quad \frac{7}{16}$$

The **Least Common denominator** is the least whole number that is a multiple of both numbers.

Step 2: Compare like fractions: the larger fraction is the one with the larger numerator:

$$\frac{10}{16} > \frac{6}{16}$$

What about this problem?

$$\frac{3}{12} \; ? \; \frac{27}{30}$$

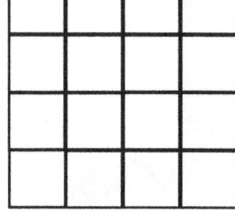

Let me think. **Step 1:** Reduce each fraction to lowest terms:

Divide both the numerator and the denominator by their greatest common factor (GCF).

$$\frac{(3 \div 3)}{(12 \div 3)} \; ? \; \frac{(27 \div 3)}{(30 \div 3)} \longrightarrow \frac{1}{4} \; ? \; \frac{9}{10}$$

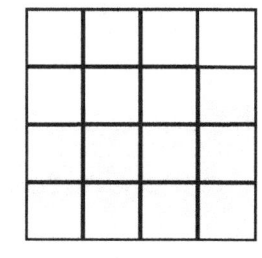

Step 2: Find a Common Denominator (change both fractions into equivalent fractions so that they both have the same denominator).

Multiples of 4: 4, 8, 12, 16, 20, 24, 28, 32, 36, 40, ...

Multiples of 10: 10, 20, 30, 40, 50, 60, 70, 80, 90, ...

The **Least Common Denominator** is 20.

What is the number that both the 4 and 10 can be divided by?

20 and 40 can be divided by 4 and 10. Here, the **least** common denominator is 20.

Step 3: 20 divided by 4 is 5, so, I need to multiply both the numerator and the denominator of $\frac{1}{4}$ by 5.

Step 4: 20 divided by 10 is 2, so, I need to multiply both the numerator and the denominator of $\frac{9}{10}$ by 2.

$$\frac{1}{4} \quad \frac{9}{10} \longrightarrow \frac{1 \times 5}{4 \times 5} \quad \frac{9 \times 2}{10 \times 2} \longrightarrow \frac{5}{20} < \frac{18}{20}$$

They are like fractions now. Compare the numerators!

$$\frac{5}{20} < \frac{18}{20}$$

since 5 is less than 18 (or you can say that 18 is greater than 5): any fractions can be compared by getting a common denominator

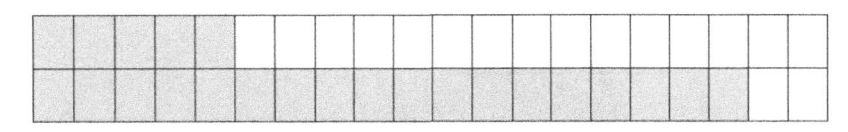

1. <u>Color</u> the correct fraction of each shape. <u>Write</u> the missing numbers. <u>Compare</u> unlike fractions using ">," "<," or "=."

$\frac{1}{6}$ — $\frac{2}{4}$ The Least Common Denominator is 12: $\frac{1 \times 2}{6 \times 2}$ — $\frac{2 \times 3}{4 \times 3}$ → $\frac{2}{12}$ — $\frac{6}{12}$

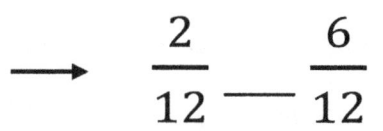

Multiples of 6: 6, 12, 18, 24, …
Multiples of 4: 4, 8, 12, 16, 20

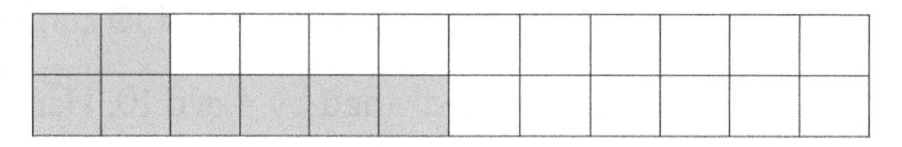

$\frac{3}{5}$ — $\frac{2}{3}$ The Least Common Denominator is _____: $\frac{3 \times 3}{5 \times 3}$ — $\frac{2 \times 5}{3 \times 5}$ → $\frac{9}{15}$ — $\frac{10}{15}$

Multiples of 5: _____
Multiples of 3: _____

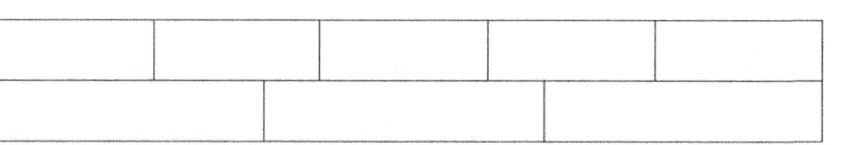

$\frac{4}{8}$ — $\frac{5}{6}$ The Least Common Denominator is _____: $\frac{4 \times}{8 \times}$ — $\frac{5 \times}{6 \times}$ → $\frac{\quad}{24}$ — $\frac{\quad}{\quad}$

Multiples of 8: _____
Multiples of 6: _____

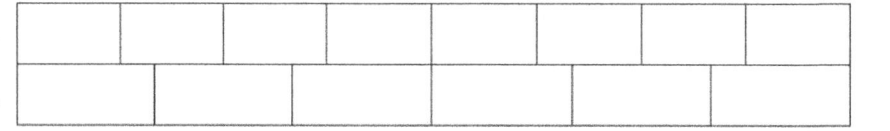

$\frac{3}{4}$ — $\frac{5}{6}$ The Least Common Denominator is _____: $\frac{\times}{\times}$ — $\frac{\times}{\times}$ → $\frac{\quad}{\quad}$ — $\frac{\quad}{\quad}$

Multiples of 4: _____
Multiples of 6: _____

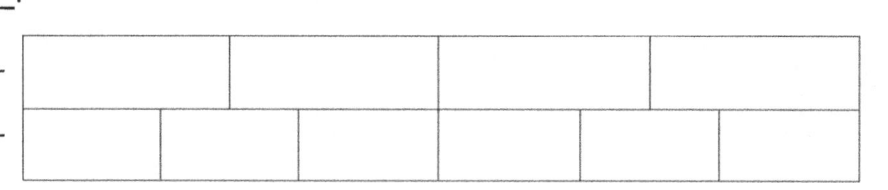

1. Color the correct fraction of each shape. Write the missing numbers. Compare unlike fractions using ">," "<," or "=."

$\dfrac{5}{6}$ — $\dfrac{7}{9}$ The Least Common Denominator is _____: $\dfrac{\times}{\times}$ — $\dfrac{\times}{\times}$ → $\dfrac{}{18}$ — $\dfrac{}{18}$

Multiples of 6: _____
Multiples of 9: _____

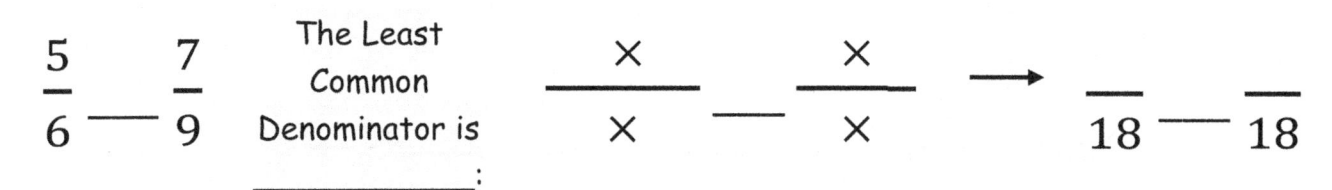

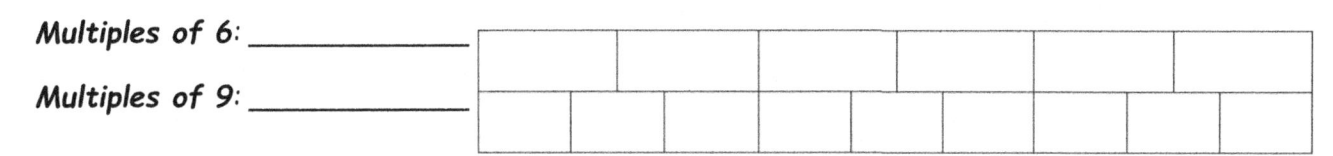

$\dfrac{2}{3}$ — $\dfrac{9}{10}$ The Least Common Denominator is _____: $\dfrac{2\times}{3\times}$ — $\dfrac{9\times}{10\times}$ → $\dfrac{}{}$ — $\dfrac{}{}$

Multiples of 3: _____
Multiples of 10: _____

2. Multiply and simplify if possible. Circle the correct answer (letter).

Find and color red $\dfrac{4}{5}$ of 20 blocks. Find and cross out $\dfrac{3}{4}$ of these red blocks.

$\dfrac{4}{5} \times 20 = \dfrac{}{} \times \dfrac{}{} = \dfrac{\times}{\times} = \underline{}$

$\dfrac{3}{4} \times \underline{} = \dfrac{}{} \times \dfrac{}{} = \dfrac{\times}{\times} = \underline{}$

a) 12

b) 16

c) 10

1. <u>Color</u> the correct fraction of each shape. <u>Write</u> the missing numbers. <u>Compare</u> unlike fractions using ">," "<," or "=."

$\dfrac{8}{9}$ —— $\dfrac{3}{4}$ Common Denominator is _____ : $\dfrac{\times}{\times}$ —— $\dfrac{\times}{\times}$ → $\dfrac{}{}$ — $\dfrac{}{}$

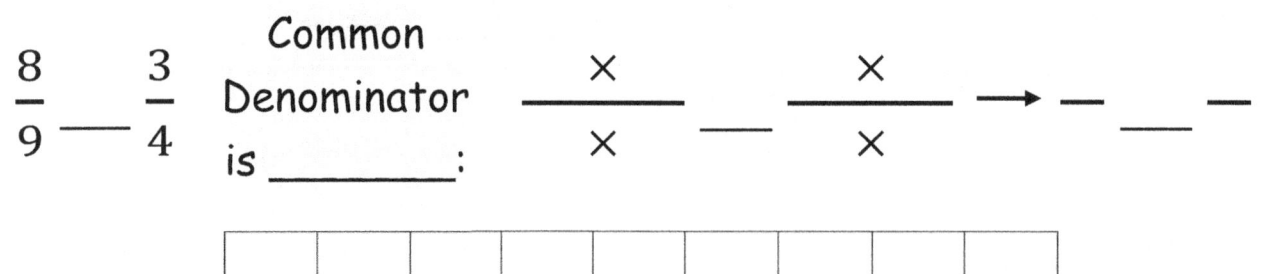

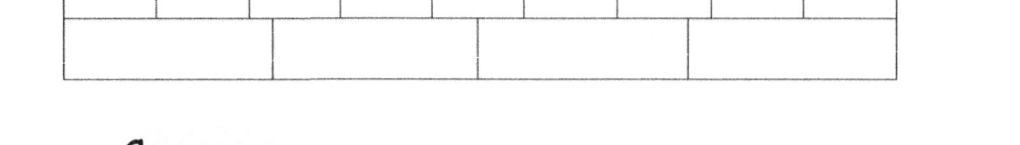

$\dfrac{5}{8}$ —— $\dfrac{7}{11}$ Common Denominator is _____ : $\dfrac{\times}{\times}$ —— $\dfrac{\times}{\times}$ → $\dfrac{}{}$ — $\dfrac{}{}$

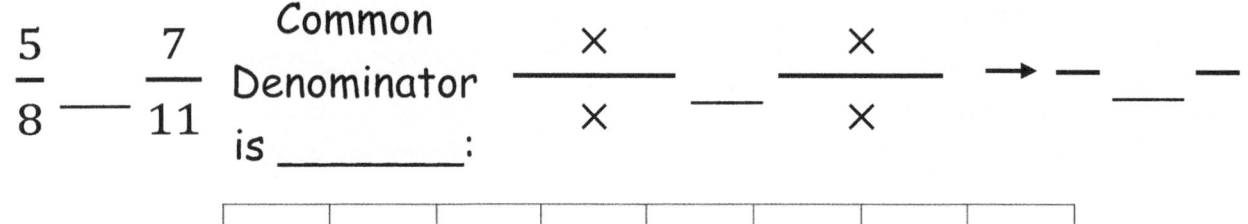

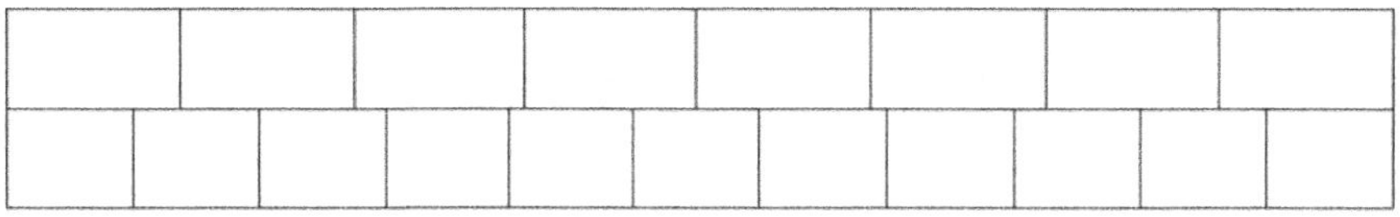

2. <u>Multiply</u> and <u>simplify</u> if possible. <u>Circle</u> the correct answer (letter).
<u>Find</u> and <u>color red</u> $\dfrac{4}{7}$ of 21 blocks. <u>Find</u> and <u>cross out</u> $\dfrac{1}{2}$ of these red blocks.

$\dfrac{4}{7} \times 21 = \dfrac{}{} \times \dfrac{}{} = \dfrac{\times}{\times} = \underline{}$

$\dfrac{1}{2} \times \underline{} = \dfrac{}{} \times \dfrac{}{} = \dfrac{\times}{\times} = \underline{}$

a) 9

b) 10

c) 6

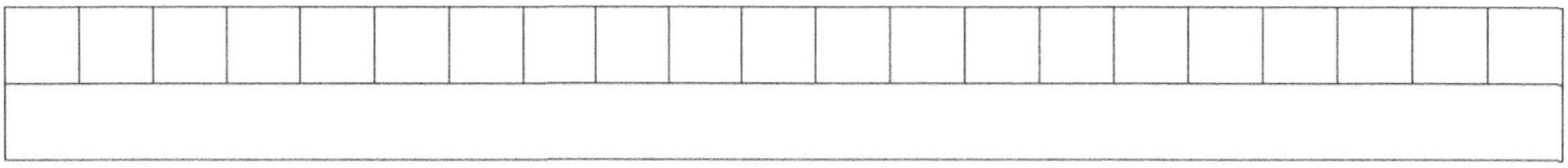

1. Color the correct fraction of each shape. Write the missing numbers. Compare unlike fractions using ">," "<," or "=."

$\dfrac{5}{9}$ ___ $\dfrac{3}{5}$ Common Denominator is _____ : $\dfrac{\times}{\times}$ ___ $\dfrac{\times}{\times}$ → $\dfrac{\ }{\ }$ ___ $\dfrac{\ }{\ }$

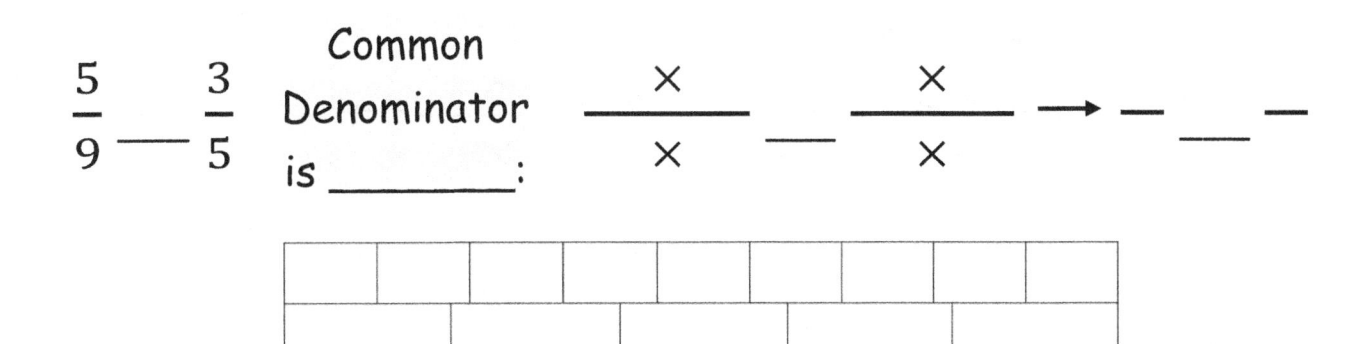

$\dfrac{4}{9}$ ___ $\dfrac{6}{10}$ Common Denominator is _____ : $\dfrac{\times}{\times}$ ___ $\dfrac{\times}{\times}$ → $\dfrac{\ }{\ }$ ___ $\dfrac{\ }{\ }$

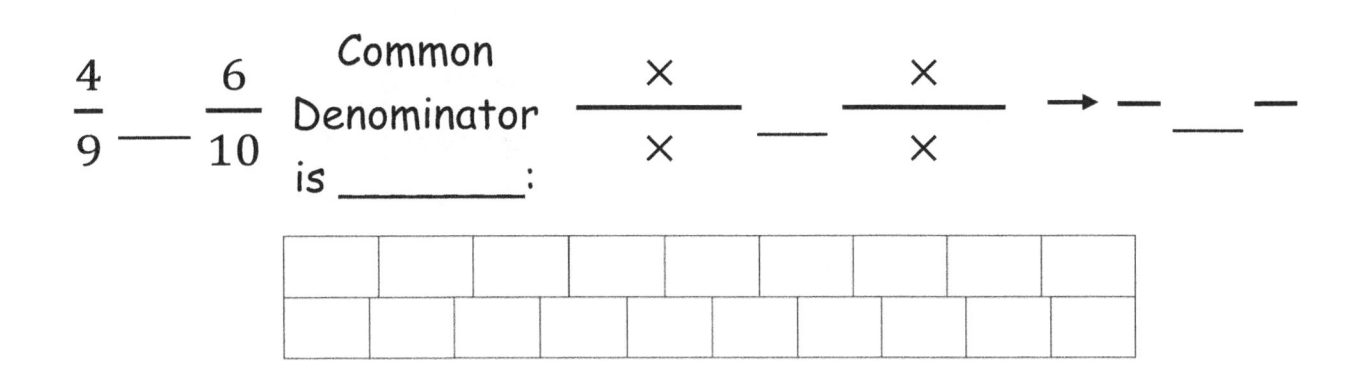

2. One fifth of a bottle can hold 200 ml of water. How much water will the bottle hold when it is $\dfrac{1}{2}$ full? Circle the correct answer.

a) 400   b) 1000   c) 750   d) 500

3. What fraction is made of flowers?

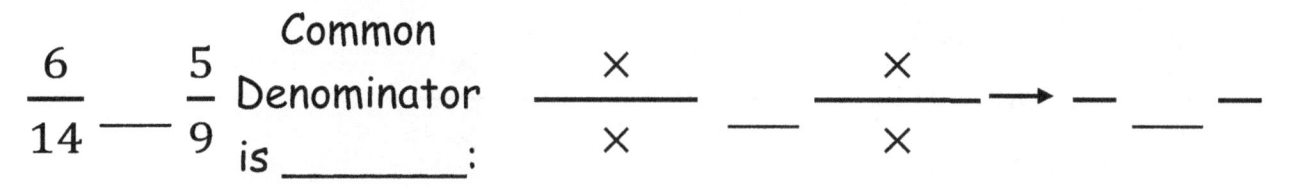

1. <u>Color</u> the correct fraction of each shape. <u>Write</u> the missing numbers. <u>Compare</u> unlike fractions using ">," "<," or "=."

$\frac{6}{14}$ —— $\frac{5}{9}$ Common Denominator is _____ : $\frac{\times}{\times}$ —— $\frac{\times}{\times}$ → $\frac{}{}$ —— $\frac{}{}$

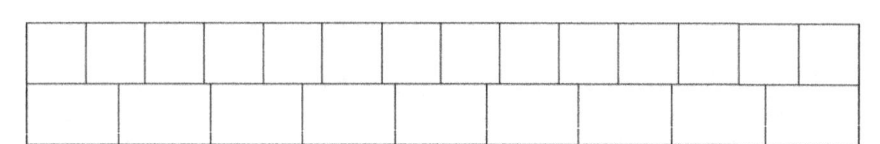

$\frac{1}{3}$ —— $\frac{3}{7}$ Common Denominator is _____ : $\frac{\times}{\times}$ —— $\frac{\times}{\times}$ → $\frac{}{}$ —— $\frac{}{}$

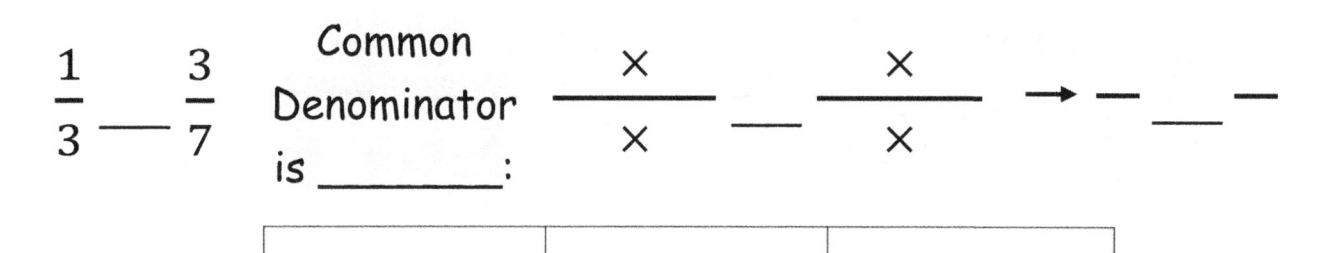

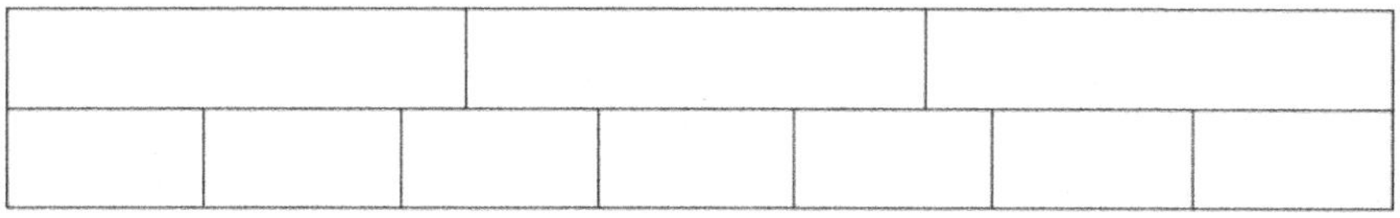

2. It takes $2\frac{1}{3}$ hours to paint a room. <u>How many hours</u> will it take to paint 3 such rooms? <u>Circle</u> the right answer.

   a) 4   b) 8   c) 6   d) 7

3. What fraction is made of stars?

1. <u>Color</u> the numerator. <u>Find</u> the Common Denominator. <u>Compare</u> unlike fractions.

I usually avoid all the tricks, but now I want to help you with the tricks since they really work.

I compare one fourth (or a quarter) and five sixths.

$$\frac{1}{4} \; ? \; \frac{5}{6}$$

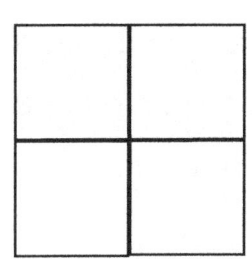

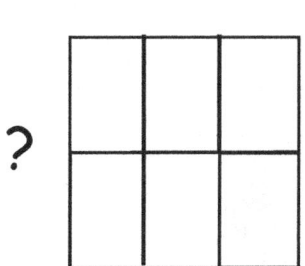

Compare the halves!

I don't see any halves!

Of course, there are no halves. But is one fourth more or less than a half?

One fourth is less than a half!

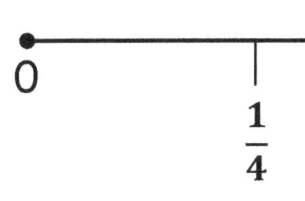

$$\frac{1}{4} < \frac{1}{2}$$

Because a half with the denominator 4 is two fourths: $\frac{2}{4}$.

And $\frac{1}{4} < \frac{2}{4}$.

Great job! Now, are five sixths more or less than a half?

Math Success Practice Workbook Fractions 4th 5th 6th Grades

Five sixths are more than a half.

$$\frac{5}{6} > \frac{1}{2}$$

since a half with the denominator 6 is three sixths: $\frac{3}{6}$. And $\frac{5}{6} > \frac{3}{6}$.

I compare one fourth (=a quarter) and five sixths through halves.

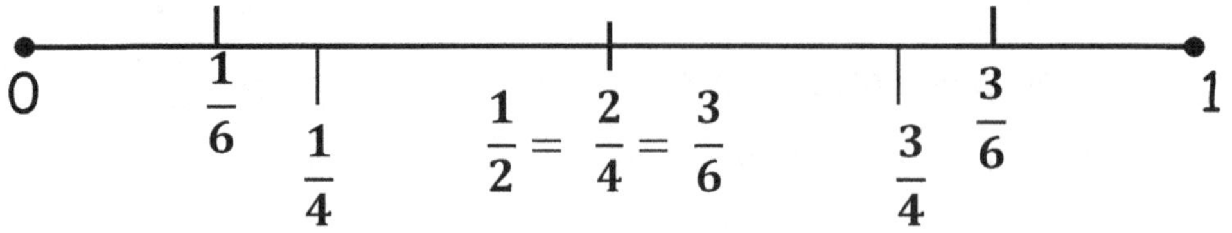

$$\frac{1}{4} < \frac{1}{2} < \frac{5}{6}$$

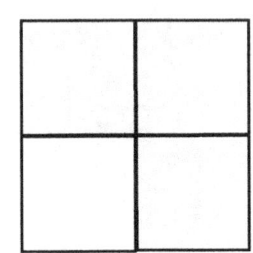

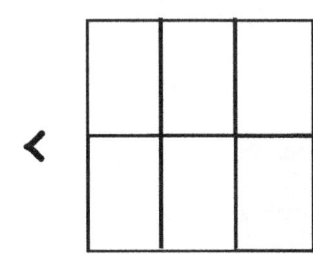

Good job, me! You're better than all the Brainers!

Back to math! How to find a half of a fraction? Look at the denominator and divide it by 2!

1. <u>Compare</u> fractions using my trick and <u>write</u> ">" or "<." <u>Color</u> the correct fraction of each shape. I'll compare $\frac{5}{6}$ and $\frac{1}{4}$.

$\frac{any\ number\ (x)}{6}$ → a half with the denominator 6 is three sixths ($\frac{3}{6}$) → compare $\frac{3}{6}$ and $\frac{5}{6}$ → $\frac{5}{6} > \frac{3}{6}$

$\frac{any\ number\ (x)}{4}$ → a half with the denominator 4 is two fourths ($\frac{2}{4}$) → compare $\frac{2}{4}$ and $\frac{1}{4}$ → $\frac{1}{4} < \frac{2}{4}$

$\frac{2}{6}$ — $\frac{7}{12}$ $\frac{2}{6}$ — $\frac{1}{2}$ — $\frac{7}{12}$

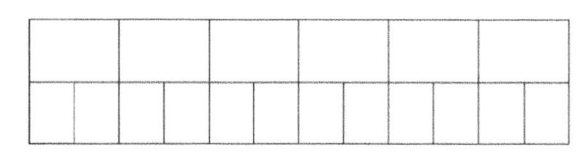

62   © 2020 Home Run Press, LLC

1. **Compare** fractions using my trick and **write** ">," "=," or "<." **Color** the correct fraction of each shape.

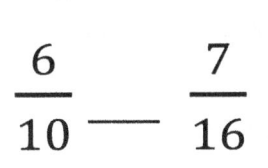

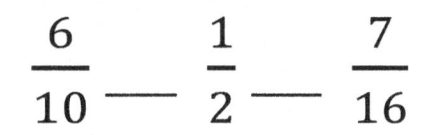

$\frac{6}{10}$ — $\frac{7}{16}$ $\frac{6}{10}$ — $\frac{1}{2}$ — $\frac{7}{16}$

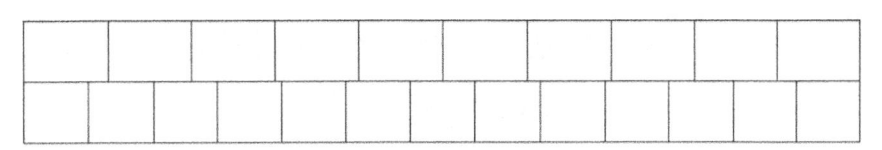

2. **Answer** the questions.

$\frac{3}{8}$ — $\frac{8}{11}$ $\frac{3}{8}$ — $\frac{1}{2}$ — $\frac{8}{11}$

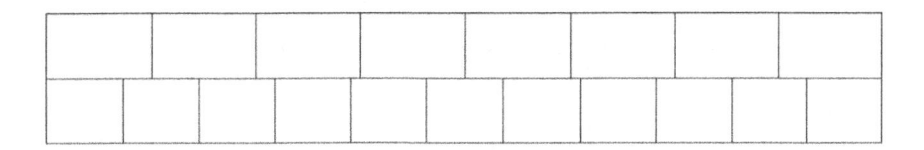

Find $\frac{1}{5}$ of flowers.

$\frac{4}{7}$ — $\frac{5}{12}$ $\frac{4}{7}$ — $\frac{1}{2}$ — $\frac{5}{12}$ Find $\frac{3}{5}$ of flowers.

$\frac{7}{9}$ — $\frac{6}{13}$ $\frac{7}{9}$ — $\frac{1}{2}$ — $\frac{6}{13}$ Find $\frac{2}{5}$ of flowers.

$\frac{3}{10}$ — $\frac{11}{18}$ $\frac{3}{10}$ — $\frac{1}{2}$ — $\frac{11}{18}$

$\frac{4}{12}$ — $\frac{6}{9}$ $\frac{4}{12}$ — $\frac{1}{2}$ — $\frac{6}{9}$

$\frac{4}{16}$ — $\frac{10}{15}$ $\frac{4}{16}$ — $\frac{1}{2}$ — $\frac{10}{15}$

What's the difference between a proper fraction and an improper fraction?

That's easy! An improper fraction always misbehaves😊! Just like me!

A proper fraction is less than a whole: that means the numerator (top number) is less than the denominator (bottom number):

a whole > $\frac{1}{2}$ (a half);

a whole > $\frac{3}{4}$ (three fourths).

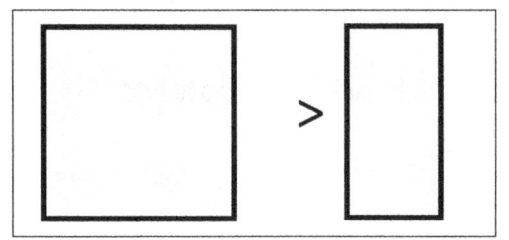

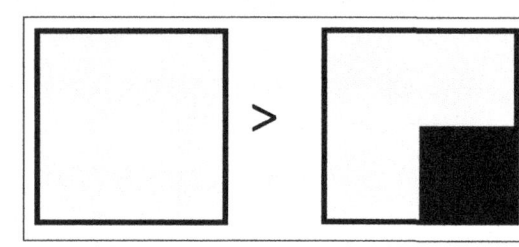

Here, an improper fraction is equal to or more than a whole: that means the numerator (top number) is equal to or greater than the denominator (bottom number):

$\frac{4}{4}$ (four fourths) = a whole

$\frac{3}{2}$ (three halves) > a whole.

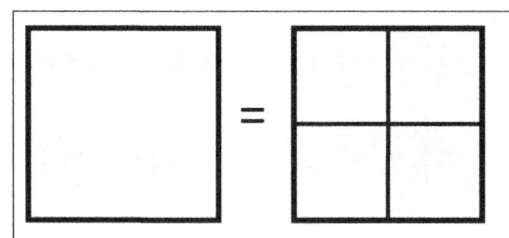

An improper fraction has a value equal to or greater than 1.

3 units of a half are greater than 2 units of a half: $\frac{3}{2}$ > $\frac{2}{2}$.

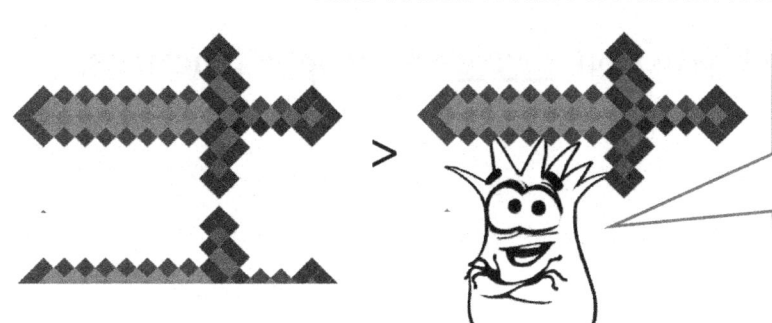

> I would say that we compare 1 whole plus $\frac{1}{2}$ and 1 whole.

Numbers like 1, 2, 3, 4, etc. are called whole numbers

You mentioned a mixed fractions. A combination of a whole number with a fraction is called a mixed number. $1\frac{1}{2}$ or $3\frac{2}{5}$ are called mixed numbers.

Hint: A mixed number = the sum of a whole number and a fraction: $1\frac{1}{2} = 1 + \frac{1}{2}$

1. <u>Cross out</u> proper fractions and <u>circle</u> improper fractions. <u>Simplify</u> if possible. Be quick! I need to build a house!

$\frac{8}{3} = —$ $\qquad$ $\frac{5}{9} = —$ $\qquad$ $\frac{4}{8} = —$ $\qquad$ $\frac{6}{2} = —$

$\frac{1}{12} = —$ $\qquad$ $\frac{4}{3} = —$ $\qquad$ $\frac{15}{16} = —$ $\qquad$ $\frac{20}{11} = —$

$\frac{7}{10} = —$ $\qquad$ $\frac{3}{8} = —$ $\qquad$ $\frac{21}{16} = —$ $\qquad$ $\frac{18}{9} = —$

$\frac{3}{9} = —$ $\qquad$ $\frac{14}{3} = —$ $\qquad$ $\frac{1}{5} = —$ $\qquad$ $\frac{3}{6} = —$

$\frac{7}{5} = —$ $\qquad$ $\frac{6}{8} = —$ $\qquad$ $\frac{3}{24} = —$ $\qquad$ $\frac{36}{12} = —$

1. <u>Draw</u> the swords to match each fraction. <u>Circle</u> the proper fractions.

$$\frac{4}{2}$$

$$\frac{2}{2}$$

2. <u>Color</u> the blocks black to match each fraction. <u>Circle</u> the improper fraction(s). <u>Cross out</u> the block(s) you do not need.

$\frac{1}{4}$

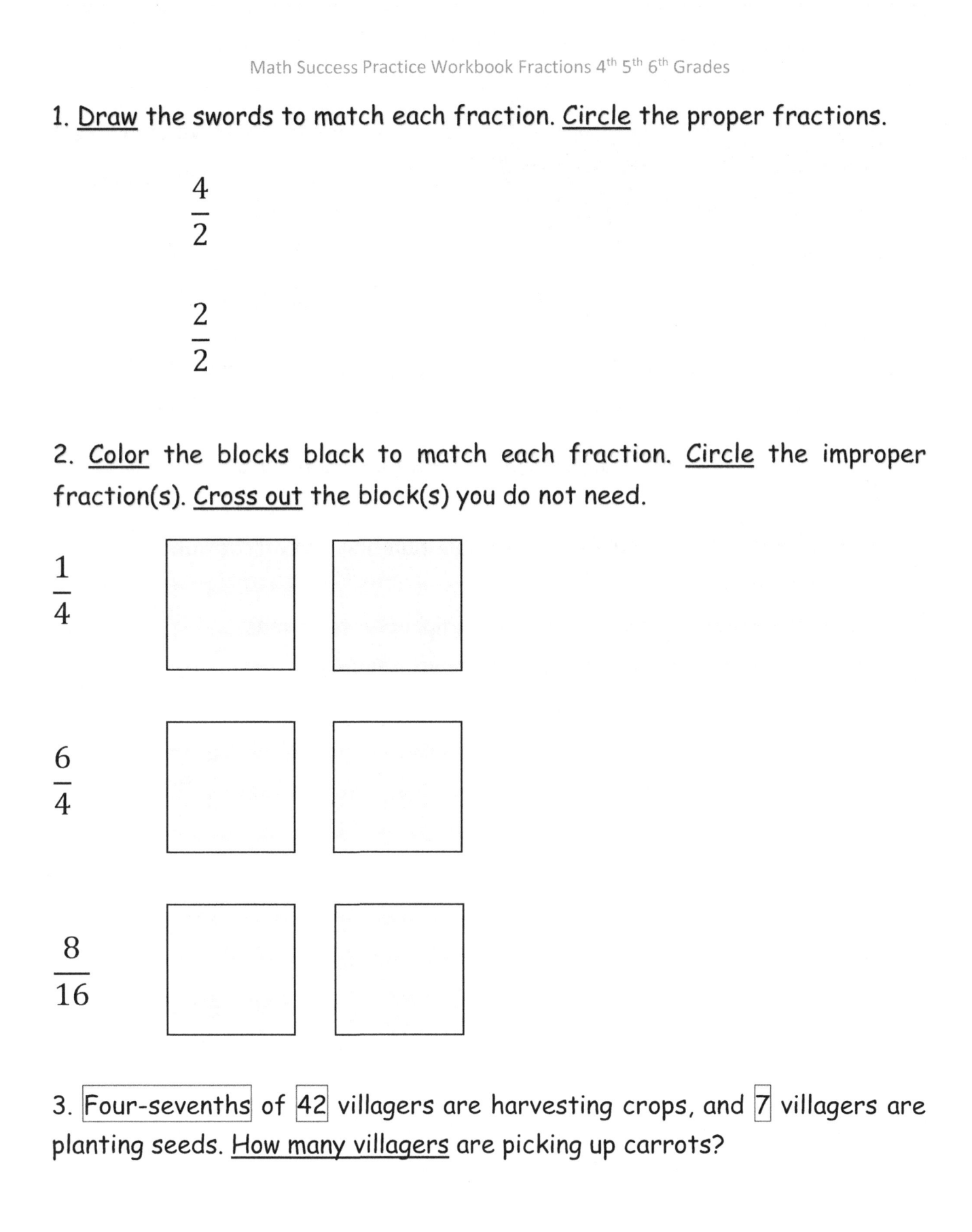

$\frac{6}{4}$

$\frac{8}{16}$

3. Four-sevenths of 42 villagers are harvesting crops, and 7 villagers are planting seeds. <u>How many villagers</u> are picking up carrots?

Answer: _____

How can I convert an improper fraction to a mixed number? For example, $\frac{3}{2}$ equals What?

Hint: The improper fraction equals the mixed number: $\frac{3}{2} = 1\frac{1}{2}$

Aha... The mixed number ($1\frac{1}{2}$) is the sum of a whole number and a fraction: $1\frac{1}{2} = 1 + \frac{1}{2}$ (a mixed number consists of a whole number part and a fraction part.)

Step 1: Ask a question: The improper fraction equals what mixed number? For example, the improper fraction $\frac{3}{2}$ equals what mixed number? $\frac{2}{2}$ is a whole. Then, $\frac{3}{2} - \frac{2}{2} = \frac{1}{2}$.

So, $\frac{3}{2} = \frac{2}{2} + \frac{1}{2} = 1 + \frac{1}{2} = 1\frac{1}{2}$

Another algorithm:

$\frac{5}{2}$ = ? mixed number

The whole number part → 2
The denominator of a fraction part → 2) 5
                                      -4
                                      ───
                                       1 ← The numerator of a fraction part

To convert an improper fraction to a mixed number, you need to divide the numerator by the denominator to get the whole number 2 and the remainder 1.

The whole number in the quotient is the whole part of your mixed number.

The remainder is the numerator of a fraction part of your mixed number.

$\frac{11}{3}$ = ? mixed number

denominator → 3) 1 1 ← = numerator of an improper fraction
                   9
                  ───
                   2 ← = remainder = the new numerator of a fraction part of a mixed number

3 ← = the whole number part of a mixed number

$\frac{11}{3} = 3\frac{2}{3}$

1. <u>Convert</u> an improper fraction to a mixed number. <u>Color</u> the blocks to match each fraction.

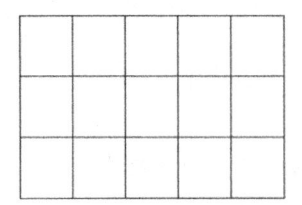

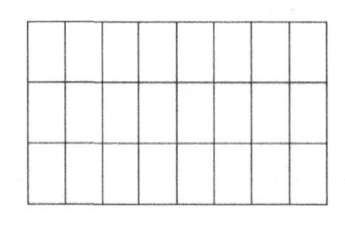

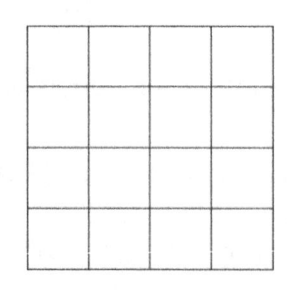

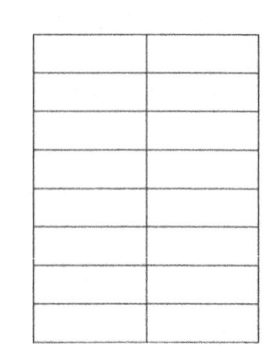

$\dfrac{11}{5} = \underline{}\dfrac{}{}$ $\dfrac{19}{8} = \underline{}\dfrac{}{}$ $\dfrac{13}{4} = \underline{}\dfrac{}{}$ $\dfrac{15}{2} = \underline{}\dfrac{}{}$

2. Two-eighths of 72 kids are boys, and four-ninths of the rest of the kids are girls. <u>How many kids</u> are girls?

Answer: _____

3. <u>Convert</u> an improper fraction to a mixed number.

$\dfrac{8}{3} = \underline{}\dfrac{}{}$ $\dfrac{9}{5} = \underline{}\dfrac{}{}$ $\dfrac{8}{4} = \underline{}\dfrac{}{}$ $\dfrac{6}{2} = \underline{}\dfrac{}{}$

$\dfrac{12}{5} = \underline{}\dfrac{}{}$ $\dfrac{10}{3} = \underline{}\dfrac{}{}$ $\dfrac{16}{8} = \underline{}\dfrac{}{}$ $\dfrac{20}{11} = \underline{}\dfrac{}{}$

$\dfrac{18}{10} = \underline{}\dfrac{}{}$ $\dfrac{24}{7} = \underline{}\dfrac{}{}$ $\dfrac{20}{16} = \underline{}\dfrac{}{}$ $\dfrac{18}{9} = \underline{}\dfrac{}{}$

1. <u>Convert</u> an improper fraction to a mixed number.

$\frac{9}{3} = \underline{}\,\frac{\underline{}}{\underline{}}$ $\qquad$ $\frac{14}{3} = \underline{}\,\frac{\underline{}}{\underline{}}$ $\qquad$ $\frac{5}{2} = \underline{}\,\frac{\underline{}}{\underline{}}$ $\qquad$ $\frac{7}{6} = \underline{}\,\frac{\underline{}}{\underline{}}$

$\frac{7}{5} = \underline{}\,\frac{\underline{}}{\underline{}}$ $\qquad$ $\frac{8}{5} = \underline{}\,\frac{\underline{}}{\underline{}}$ $\qquad$ $\frac{24}{3} = \underline{}\,\frac{\underline{}}{\underline{}}$ $\qquad$ $\frac{36}{12} = \underline{}\,\frac{\underline{}}{\underline{}}$

$\frac{22}{3} = \underline{}\,\frac{\underline{}}{\underline{}}$ $\qquad$ $\frac{34}{6} = \underline{}\,\frac{\underline{}}{\underline{}}$ $\qquad$ $\frac{11}{2} = \underline{}\,\frac{\underline{}}{\underline{}}$ $\qquad$ $\frac{28}{5} = \underline{}\,\frac{\underline{}}{\underline{}}$

$\frac{42}{9} = \underline{}\,\frac{\underline{}}{\underline{}}$ $\qquad$ $\frac{25}{4} = \underline{}\,\frac{\underline{}}{\underline{}}$ $\qquad$ $\frac{63}{8} = \underline{}\,\frac{\underline{}}{\underline{}}$ $\qquad$ $\frac{36}{10} = \underline{}\,\frac{\underline{}}{\underline{}}$

$\frac{31}{7} = \underline{}\,\frac{\underline{}}{\underline{}}$ $\qquad$ $\frac{46}{5} = \underline{}\,\frac{\underline{}}{\underline{}}$ $\qquad$ $\frac{25}{12} = \underline{}\,\frac{\underline{}}{\underline{}}$ $\qquad$ $\frac{41}{15} = \underline{}\,\frac{\underline{}}{\underline{}}$

2. Two-sevenths of 35 fruits are apples and the rest are oranges. <u>How many more or less apples than oranges</u> are there?

Answer: _____

3. I have run away from five-eighths of the 32 zombies I had met in my game. My brother run away from four-ninths of the 27 zombies he had met in his game. <u>Who</u> was luckier and run away from the greatest number of zombies?

Answer: _____

How can I convert a mixed number to an improper fraction?

For example, $2\frac{2}{3}$ equals What?

Hint: The mixed number equals the improper fraction:

$2\frac{1}{2}$ (two swords and a half) = $\frac{5}{2}$ (five halves of swords)

Correct. Let's think!

Step 1: The whole number part equals what fraction? The whole number part of 1 equals: $1 = \frac{2}{2}$. The whole number part of 2 equals: $2 = \frac{4}{2}$.

Step 2: Add the two fractions: $\frac{4}{2} + \frac{1}{2} = \frac{4+1}{2} = \frac{5}{2}$

Another algorithm: $2\frac{1}{2} = ?$

Step 1: Multiply the whole number part by the denominator (in our problem we multiply 2 by 2).

→ 2 × 2 = 4 (the product is the numerator of the whole number part)

Step 2: Add the product to the numerator of a fraction part (in our problem it's 1).

→ 4 + 1 = 5 (we added the numerator of the whole number part and the numerator of the fraction part that equals the numerator of the improper fraction)

$$2\frac{1}{2} = \frac{5}{2}$$

1. **Convert** a mixed number to an improper fraction.

$1\frac{3}{5} = \dfrac{}{}$	$2\frac{5}{8} = \dfrac{}{}$	$3\frac{1}{4} = \dfrac{}{}$

1) 1 × 5 = ___	1) __ × __ = ___	1) __ × __ = ___

2) __ + 3 = ___	2) __ + __ = ___	2) __ + __ = ___

2. Three-tenths of 50 blooming flowers were blue orchid. If sunflower were one-tenths more than blue orchid, and the rest of the flowers were divided by peonies and lilacs, and roses respectively, how many lilacs were blooming?

Answer: _____

3. **Convert** a mixed number to an improper fraction. **Write** the missing numbers.

$4\frac{2}{5} = \dfrac{(4\times 5)+2}{5} = \dfrac{}{}$	$7\frac{1}{2} = \dfrac{(\_\times\_)+\_}{\_} = \dfrac{}{}$

$2\frac{1}{4} = \dfrac{(\_\times\_)+\_}{\_} = \dfrac{}{}$	$6\frac{2}{3} = \dfrac{(\_\times\_)+\_}{\_} = \dfrac{}{}$

$3\frac{1}{2} = \dfrac{(\_\times\_)+\_}{\_} = \dfrac{}{}$	$3\frac{5}{8} = \dfrac{(\_\times\_)+\_}{\_} = \dfrac{}{}$

1. <u>Convert</u> a mixed number to an improper fraction. <u>Write</u> the missing numbers.

$7\dfrac{3}{4} = \dfrac{(\underline{} \times \underline{}) + \underline{}}{\underline{}} = \dfrac{}{}$ $\qquad$ $9\dfrac{4}{5} = \dfrac{(\underline{} \times \underline{}) + \underline{}}{\underline{}} = \dfrac{}{}$

$3\dfrac{3}{8} = \dfrac{(\underline{} \times \underline{}) + \underline{}}{\underline{}} = \dfrac{}{}$ $\qquad$ $5\dfrac{5}{8} = \dfrac{(\underline{} \times \underline{}) + \underline{}}{\underline{}} = \dfrac{}{}$

$6\dfrac{1}{10} = \dfrac{(\underline{} \times \underline{}) + \underline{}}{\underline{}} = \dfrac{}{}$ $\qquad$ $3\dfrac{2}{9} = \dfrac{(\underline{} \times \underline{}) + \underline{}}{\underline{}} = \dfrac{}{}$

$10\dfrac{3}{5} = \dfrac{}{} = \dfrac{}{}$ $\qquad$ $7\dfrac{9}{11} = \dfrac{}{} = \dfrac{}{}$

2. I met some horses of two colors. Three-sixths of the horses were white. 24 horses were black. <u>How many horses</u> did I meet?

Answer: _____

3. The villager planted some melon and pumpkin seeds. Two-fifths of them were melon seeds. If he planted 45 pumpkin seeds, <u>how many melon and pumpkin seeds</u> did he have at first?

Answer: _____

1. <u>Convert</u> an improper fraction to a mixed number. <u>Circle</u> the correct answer.

$$\frac{9}{2} = \_\_ \ \_\_ \qquad \frac{25}{8} = \_\_ \ \_\_ \qquad \frac{31}{4} = \_\_ \ \_\_ \qquad \frac{37}{5} = \_\_ \ \_\_$$

a) $5\frac{2}{2}$ a) $2\frac{5}{8}$ a) $5\frac{3}{4}$ a) $7\frac{2}{5}$

b) $1\frac{3}{2}$ b) $3\frac{1}{8}$ b) $8\frac{1}{4}$ b) $6\frac{7}{5}$

c) $4\frac{1}{2}$ c) $5\frac{1}{8}$ c) $4\frac{3}{4}$ c) $5\frac{21}{5}$

d) $4\frac{2}{2}$ d) $3\frac{2}{8}$ d) $7\frac{3}{4}$ d) $4\frac{4}{5}$

2. I am an improper fraction. The sum of my numerator and denominator is 14, their difference is 2^3 (2 × 2 × 2). <u>What fraction</u> am I?

a) $\frac{8}{6}$ b) $\frac{15}{1}$ c) $\frac{6}{8}$ d) $\frac{11}{3}$

3. <u>True or false?</u> <u>Explain.</u> <u>Circle</u> the correct answer.

To convert an improper fraction to a mixed number, you need to divide the denominator by the numerator.

_____

_____

a)   TRUE   b) FALSE

1. <u>Convert</u> a mixed number to an improper fraction. <u>Circle</u> the correct answer.

$2\frac{1}{4} = \_\_\_$ $7\frac{3}{8} = \_\_\_$ $5\frac{7}{12} = \_\_\_$ $3\frac{2}{5} = \_\_\_$

a) $\frac{8}{4}$ a) $\frac{60}{8}$ a) $\frac{66}{12}$ a) $\frac{17}{5}$

b) $\frac{3}{4}$ b) $\frac{56}{8}$ b) $\frac{70}{12}$ b) $\frac{10}{5}$

c) $\frac{9}{4}$ c) $\frac{58}{8}$ c) $\frac{67}{12}$ c) $\frac{21}{5}$

d) $\frac{10}{4}$ d) $\frac{59}{8}$ d) $\frac{77}{12}$ d) $\frac{16}{5}$

2. I am an [improper] fraction. The product of my numerator and denominator is [27], their quotient is [3] <u>What fraction</u> am I?

a) $\frac{27}{9}$ b) $\frac{15}{14}$ c) $\frac{12}{4}$ d) $\frac{9}{3}$

3. <u>True or false?</u> <u>Explain.</u> <u>Circle</u> the correct answer.

The smaller the denominator the greater the fraction.

_____

_____

   b)   TRUE   b) FALSE

Math Success Practice Workbook Fractions 4th 5th 6th Grades

I need to add fractions:

$$\frac{1}{4} + \frac{5}{6} = ?$$

That's impossible!

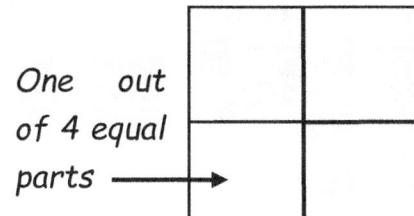

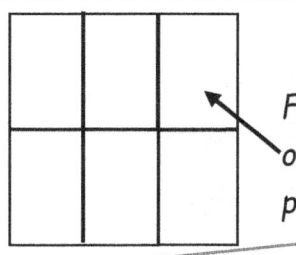

One out of 4 equal parts

Five out of 6 equal parts

Just imagine, I ate $\frac{1}{4}$ of a Strawberry Chocolate bar and $\frac{3}{8}$ of a Hazelnut Chocolate bar. <u>How much</u> did I eat?

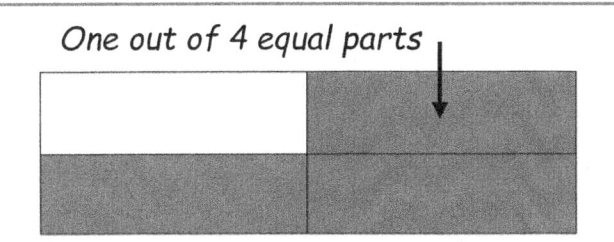

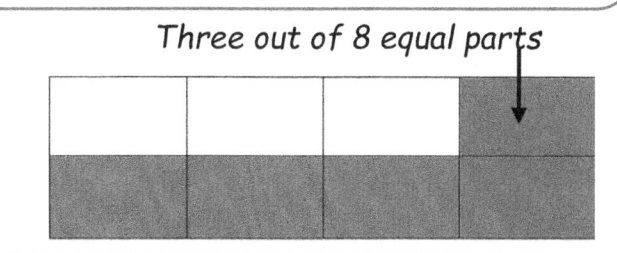

One out of 4 equal parts

Three out of 8 equal parts

Step 1: Find out equivalent fractions to get the same common denominator (the common denominator is LCD)

Step 2: Add the numerators only

Step 3: Simplify where possible

Yeah, first, we need to find out equivalent fractions!

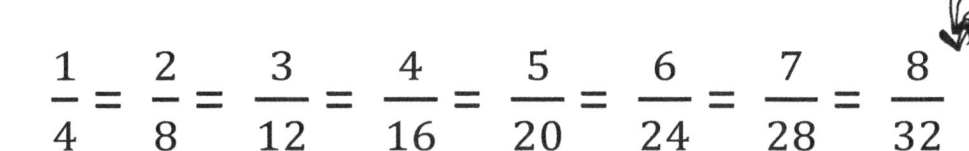

$$\frac{1}{4} = \frac{2}{8} = \frac{3}{12} = \frac{4}{16} = \frac{5}{20} = \frac{6}{24} = \frac{7}{28} = \frac{8}{32}$$

$$\frac{3}{8} = \frac{6}{16} = \frac{9}{24} = \frac{12}{32}$$

Equivalent fractions are found out by multiplying both the numerator and denominator by the same number

© 2020 Home Run Press, LLC

I have several like fractions with 8, 16, 24, 32 in the denominator. I colored the fractions strips grey.

$$\frac{1}{4} = \frac{2}{8} = \frac{3}{12} = \frac{4}{16} = \frac{5}{20} = \frac{6}{24} = \frac{7}{28} = \frac{8}{32}$$

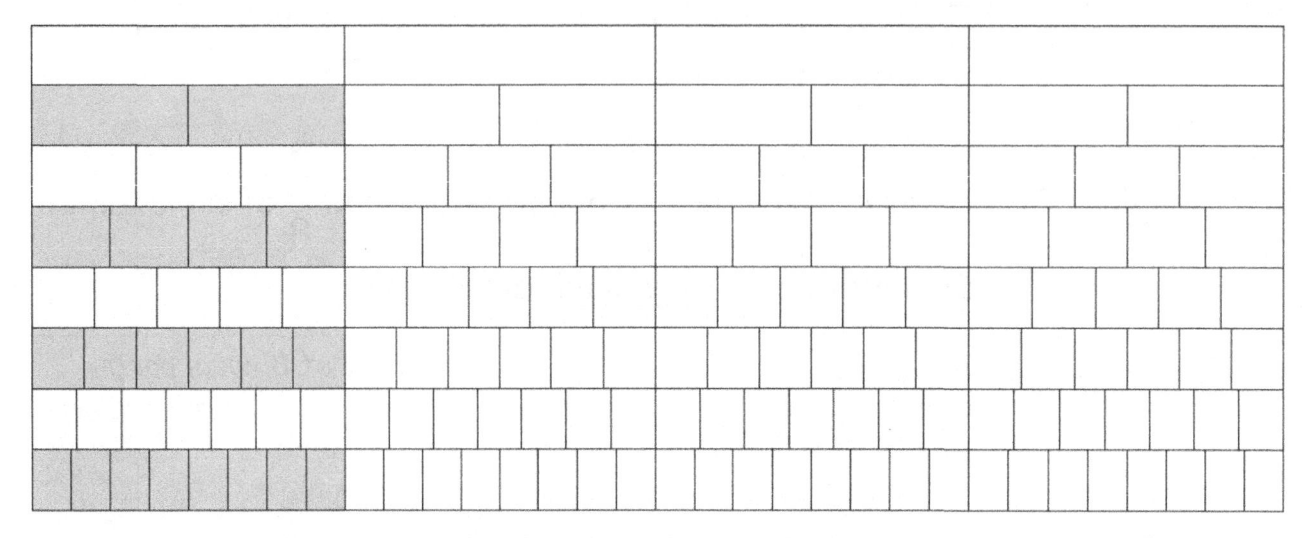

1 strip out of 4 strips is the same as 2 strips out of 8 strips, or 3 strips out of 12 strips, or 4 strips out of 16 strips, or 5 strips out of 20 strips, or 6 strips out of 24 strips, or 7 strips out of 28 strips, or 8 strips out of 32 strips – their value remains the same

$$\frac{3}{8} = \frac{6}{16} = \frac{9}{24} = \frac{12}{32}$$

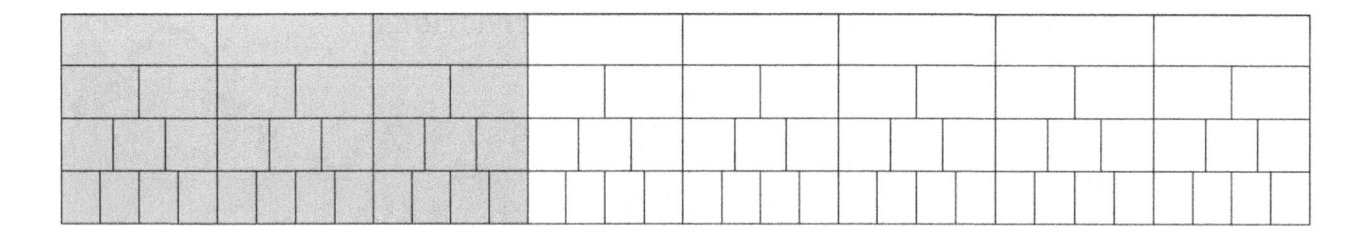

3 strips out of 8 strips are the same as 6 strips out of 16 strips, or 9 strips out of 24 strips, or 12 strips out of 32 strips – their value remains the same

Aha! The denominators are the same in these fractions! These are common denominators.

Hint: *The denominator (8) is the numeral that tells the number of parts a whole is divided into*

$$\frac{2}{8} = \frac{4}{16} = \frac{6}{24} = \frac{8}{32}$$

$$\frac{3}{8} = \frac{6}{16} = \frac{9}{24} = \frac{12}{32}$$

Adding fractions is possible ONLY with first writing the fractions with common denominators

Step 1 is done! We changed the given fractions to equivalent fractions to find common denominators. Now we must find the least common denominator.

8 is the LEAST common denominator for

$$\frac{1}{4} \ (= \frac{2}{8}) \text{ and } \frac{3}{8}.$$

These are like fractions since they both have 8 in the denominator.

So, you can write 8 as a common denominator for both fractions.

And then, add the numerators 2 and 3 to find the sum.

Step 1: Find the Common Denominator

Step 2: Add the Numerators to find the sum

$$\frac{1}{4} + \frac{3}{8} = \frac{2}{8} + \frac{3}{8} = \frac{2+3}{8} = \frac{5}{8}$$

My problem is:

$$\frac{1}{4} + \frac{5}{6} = ?$$

Don't worry, we'll start slow, I may have some ideas. Find the multiples!

Hint: To find the multiples of a number, multiply this number by the counting numbers (1, 2, 3, 4, etc.) 😊

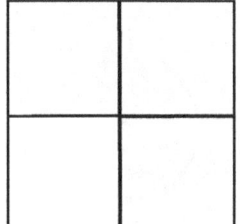

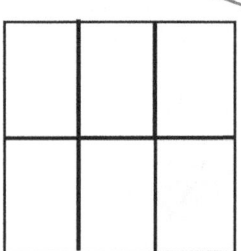

Step 1: Find the LEAST common multiple of 4 and 6.

Multiples of 4

For the denominator 4: $1 \times 4 = 4$ $2 \times 4 = 8$ $3 \times 4 = 12$

For the denominator 6: $1 \times 6 = 4$ $2 \times 6 = 12$ $3 \times 6 = 18$

Multiples of 6

The LEAST common multiple of 4 and 6 is 12:

$1 \times 4 = 4$ $2 \times 4 = 8$ $3 \times 4 = \boxed{12}$

$1 \times 6 = 4$ $2 \times 6 = \boxed{12}$ $3 \times 6 = 18$

How to find the least common multiple

Step 1: List all the multiples of the first denominator

Step 2: List all the multiples of the second denominator

Step 3: Circle the least multiple which is common to both denominators

Step 2: Draw a fractions bar, write 12 for the common denominator, and draw a slash above the numerator of each fraction:

$$\frac{1}{4} + \frac{5}{6} = \frac{\_ + \_}{12} =$$

Step 3: Divide the least common multiple by each denominator and write the quotient above each slash:

Hint: Use slashes to make calculations easy and fun 😊

$$\overset{3}{\cancel{}}\frac{1}{4} + \overset{2}{\cancel{}}\frac{5}{6} = \frac{\_ + \_}{12} =$$

$$12 \div 4 = 3$$
$$12 \div 6 = 2$$

Step 4: Multiply each numerator by the number above a slash:

$$3 \times 1 = 3$$
$$2 \times 5 = 10$$

$$\overset{3}{\cancel{}}\frac{1}{4} + \overset{2}{\cancel{}}\frac{5}{6} = \frac{(3 \times 1) + (2 \times 5)}{12} =$$

Hint: To add fractions, make sure that you add like fractions. That means they must have a common denominator 😊

You got the "new" numerators.

These are the numerators for the denominator 12:

$$\frac{1}{4} = \frac{3}{12}; \frac{5}{6} = \frac{10}{12}.$$

Step 5: Add the "new" numerators:

$$\overset{3}{\cancel{}}\frac{1}{4} + \overset{2}{\cancel{}}\frac{5}{6} = \frac{3 + 10}{12} = \frac{13}{12}$$

1. <u>Add</u> fractions. <u>Write</u> the missing numbers. <u>Color</u> the fraction of each shape. <u>Simplify</u> if possible.

$\frac{1}{4} + \frac{1}{3} = ?$ $\qquad\qquad\qquad\qquad$ $\frac{3}{4} + \frac{2}{5} = ?$

Step 1: Find the least common denominator

$\frac{1}{4} = \frac{2}{\_\_} = \frac{3}{\_\_}$ $\qquad\qquad$ $\frac{3}{4} = \frac{6}{\_\_} = \frac{9}{\_\_} = \frac{12}{\_\_} = \frac{15}{\_\_}$

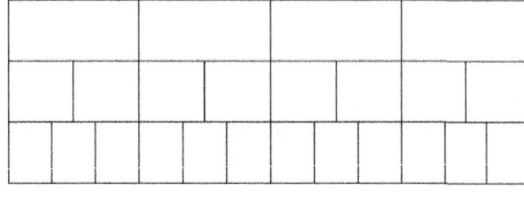

$\frac{1}{3} = \frac{2}{\_\_} = \frac{3}{\_\_} = \frac{4}{\_\_}$

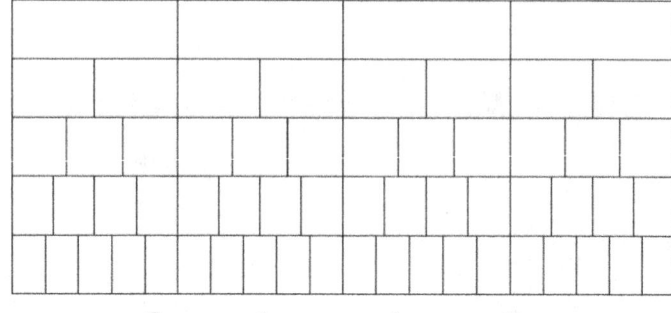

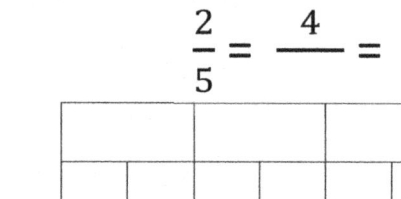

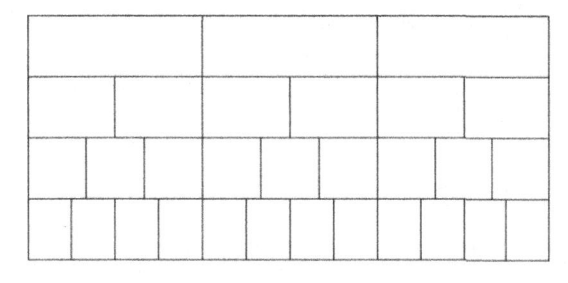

$\frac{2}{5} = \frac{4}{\_\_} = \frac{6}{\_\_} = \frac{8}{\_\_}$

___ is the least common denominator

Step 2: Add the Numerators to find the sum $\qquad$ ___ is the least common denominator

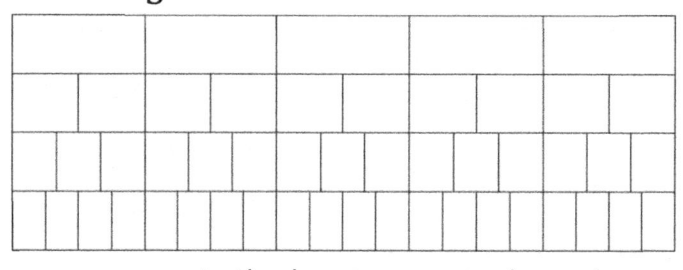

$\frac{1}{4} + \frac{1}{3} = \frac{\_\_ + \_\_}{\_\_} = \frac{\_\_}{\_\_} = \frac{\_\_}{\_\_}$ $\qquad$ $\frac{3}{4} + \frac{2}{5} = \frac{\_\_ + \_\_}{\_\_} = \frac{\_\_}{\_\_} = \frac{\_\_}{\_\_}$

2. <u>True or false?</u> <u>Explain</u>. <u>Circle</u> the correct answer.

The greater the numerator in like fractions the smaller the fraction.

_____

a)   TRUE $\qquad\qquad\qquad\qquad$ b) FALSE

80

1. <u>Add</u> fractions. <u>Write</u> the missing numbers. <u>Color</u> the fraction of each shape. <u>Simplify</u> if possible.

$$\frac{2}{3} + \frac{5}{12} = ? \qquad\qquad \frac{5}{6} + \frac{3}{4} = ?$$

Step 1: Find the least common denominator

$$\frac{2}{3} = \frac{4}{\underline{}} = \frac{6}{\underline{}} = \frac{8}{\underline{}} \qquad\qquad \frac{5}{6} = \frac{10}{\underline{}}$$

$$\frac{5}{12} \qquad\qquad\qquad \frac{3}{4} = \frac{6}{\underline{}} = \frac{9}{\underline{}}$$

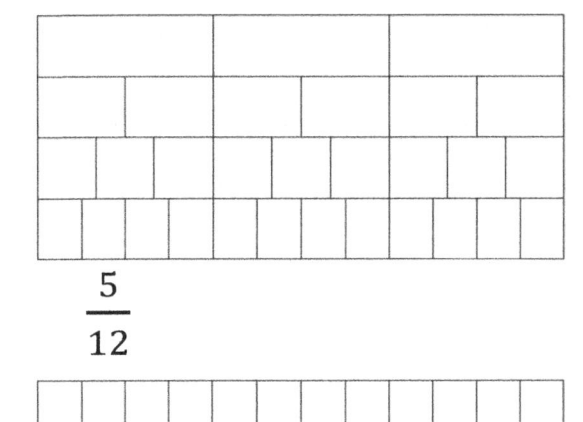

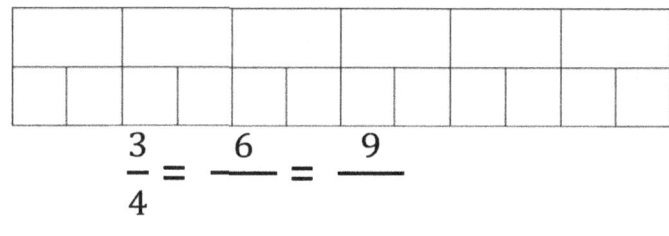

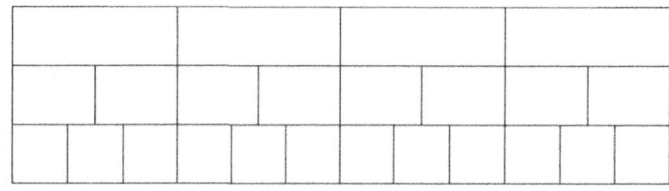

____ is the least common denominator   ____ is the least common denominator

Step 2: Add the Numerators to find the sum

$$\frac{2}{3} + \frac{5}{12} = \frac{+}{\underline{}} = \frac{-}{-} = \frac{-}{-} \qquad\qquad \frac{5}{6} + \frac{3}{4} = \frac{+}{\underline{}} = \frac{-}{-} = \frac{-}{-}$$

2. <u>True or false?</u> <u>Explain.</u> <u>Circle</u> the correct answer.

The greater the denominator in unlike fractions the greater the fraction.

_____

a) TRUE b) FALSE

1. **Add** fractions. **Write** the missing numbers. **Simplify** if possible.

$\dfrac{2}{6} + \dfrac{1}{2} = \dfrac{+}{} = \dfrac{}{} = \dfrac{}{}$ $\dfrac{2}{4} + \dfrac{4}{5} = \dfrac{+}{} = \dfrac{}{} = \dfrac{}{}$

$\dfrac{1}{3} + \dfrac{1}{4} = \dfrac{+}{} = \dfrac{}{} = \dfrac{}{}$ $\dfrac{2}{3} + \dfrac{1}{6} = \dfrac{+}{} = \dfrac{}{} = \dfrac{}{}$

$\dfrac{4}{6} + \dfrac{2}{4} = \dfrac{+}{} = \dfrac{}{} = \dfrac{}{}$ $\dfrac{4}{9} + \dfrac{6}{18} = \dfrac{+}{} = \dfrac{}{} = \dfrac{}{}$

$\dfrac{5}{14} + \dfrac{3}{7} = \dfrac{+}{} = \dfrac{}{} = \dfrac{}{}$ $\dfrac{1}{6} + \dfrac{3}{8} = \dfrac{+}{} = \dfrac{}{} = \dfrac{}{}$

$\dfrac{3}{4} + \dfrac{7}{16} = \dfrac{+}{} = \dfrac{}{} = \dfrac{}{}$ $\dfrac{4}{5} + \dfrac{5}{6} = \dfrac{+}{} = \dfrac{}{} = \dfrac{}{}$

$\dfrac{7}{12} + \dfrac{7}{8} = \dfrac{+}{} = \dfrac{}{} = \dfrac{}{}$ $\dfrac{3}{10} + \dfrac{4}{6} = \dfrac{+}{} = \dfrac{}{} = \dfrac{}{}$

Hint: Find the least common denominator by multiplying the greatest given denominator by 2, then, by 3, and so on until you get a number that is exactly divided by the given denominators.

2. **True or false?** **Explain**. **Circle** the correct answer.

The greater the denominator in unlike fractions the greater the fraction.

_____

a) TRUE b) FALSE

1. Add fractions. Write the missing numbers. Simplify if possible.

$\frac{9}{16} + \frac{11}{12} = \frac{\quad + \quad}{\quad} = \frac{\quad}{\quad} = \underline{\quad}\frac{\quad}{\quad}$

$\frac{9}{18} + \frac{1}{4} = \frac{\quad + \quad}{\quad} = \frac{\quad}{\quad} = \underline{\quad}\frac{\quad}{\quad}$

$\frac{7}{20} + \frac{5}{8} = \frac{\quad + \quad}{\quad} = \frac{\quad}{\quad} = \underline{\quad}\frac{\quad}{\quad}$

$\frac{13}{18} + \frac{7}{10} = \frac{\quad + \quad}{\quad} = \frac{\quad}{\quad} = \underline{\quad}\frac{\quad}{\quad}$

$\frac{8}{25} + \frac{4}{15} = \frac{\quad + \quad}{\quad} = \frac{\quad}{\quad} = \underline{\quad}\frac{\quad}{\quad}$

$\frac{10}{28} + \frac{3}{4} = \frac{\quad + \quad}{\quad} = \frac{\quad}{\quad} = \underline{\quad}\frac{\quad}{\quad}$

$\frac{9}{20} + \frac{11}{14} + \frac{2}{7} = \frac{\quad + \quad + \quad}{\quad} = \frac{\quad}{\quad} = \underline{\quad}\frac{\quad}{\quad}$

$\frac{3}{7} + \frac{9}{14} + \frac{13}{21} = \frac{\quad + \quad + \quad}{\quad} = \frac{\quad}{\quad} = \underline{\quad}\frac{\quad}{\quad}$

$\frac{1}{2} + \frac{7}{13} + \frac{5}{26} = \frac{\quad + \quad + \quad}{\quad} = \frac{\quad}{\quad} = \underline{\quad}\frac{\quad}{\quad}$

2. True or false? Explain. Circle the correct answer.

The least common denominator is the least factor for the numerator.

_____

a) TRUE　　　　　　　　　　　　b) FALSE

1. <u>Add</u> fractions. <u>Write</u> the missing numbers. <u>Simplify</u> if possible.

$\dfrac{2}{13} + \dfrac{5}{39} = \dfrac{\quad + \quad}{\quad} = \dfrac{\quad}{\quad} = \dfrac{\quad}{\quad}$ $\dfrac{2}{5} + \dfrac{3}{10} =$

$\dfrac{5}{8} + \dfrac{5}{6} =$ $\dfrac{5}{6} + \dfrac{7}{15} =$

$\dfrac{7}{18} + \dfrac{8}{15} =$ $\dfrac{5}{18} + \dfrac{3}{12} =$

A common denominator may be found by multiplying together given denominators!

$\dfrac{11}{15} + \dfrac{7}{9} =$ $\dfrac{9}{13} + \dfrac{1}{4} =$

$\dfrac{13}{16} + \dfrac{7}{8} =$ $\dfrac{3}{21} + \dfrac{1}{6} =$

$\dfrac{5}{12} + \dfrac{3}{8} =$ $\dfrac{7}{16} + \dfrac{5}{7} =$

2. <u>True or false?</u> <u>Explain.</u> <u>Circle</u> the correct answer.

The mixed number is the sum of a proper fraction and an improper fraction.

_____

a) TRUE   b) FALSE

I need to subtract fractions:

$$\frac{3}{4} - \frac{1}{6} = \;?$$

Let's try...

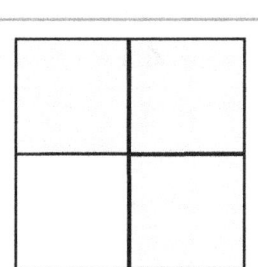

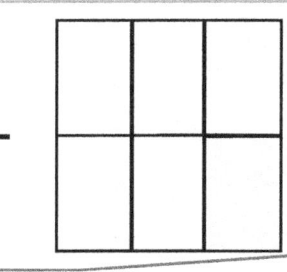

I ate $\frac{1}{6}$ out of $\frac{3}{4}$ of a Hazelnut Chocolate bar. How much is left? Can we apply the same steps to subtraction as we did in addition?

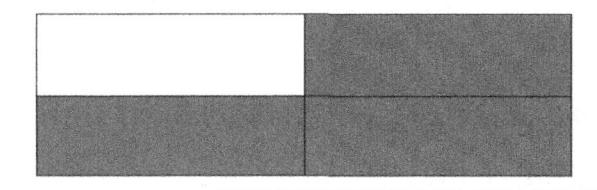

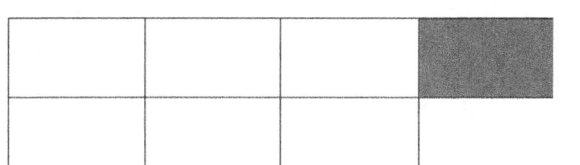

Step 1: Find out equivalent fractions to get the same common denominator (the common denominator is LCD)

Step 2: Subtract the numerators only

Step 3: Simplify where possible

Find out equivalent fractions!

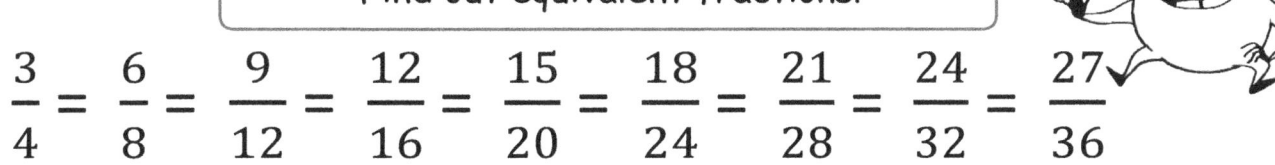

$$\frac{3}{4} = \frac{6}{8} = \frac{9}{12} = \frac{12}{16} = \frac{15}{20} = \frac{18}{24} = \frac{21}{28} = \frac{24}{32} = \frac{27}{36}$$

$$\frac{1}{6} = \frac{2}{12} = \frac{3}{18} = \frac{4}{24} = \frac{5}{30} = \frac{6}{36}$$

Equivalent fractions are found out by multiplying both the numerator and denominator by the same number

I have several like fractions with 12, 24, 36 in the denominator. I colored the fractions strips grey.

$$\frac{3}{4} = \frac{6}{8} = \frac{9}{12} = \frac{12}{16} = \frac{15}{20} = \frac{18}{24} = \frac{21}{28} = \frac{24}{32} = \frac{27}{36}$$

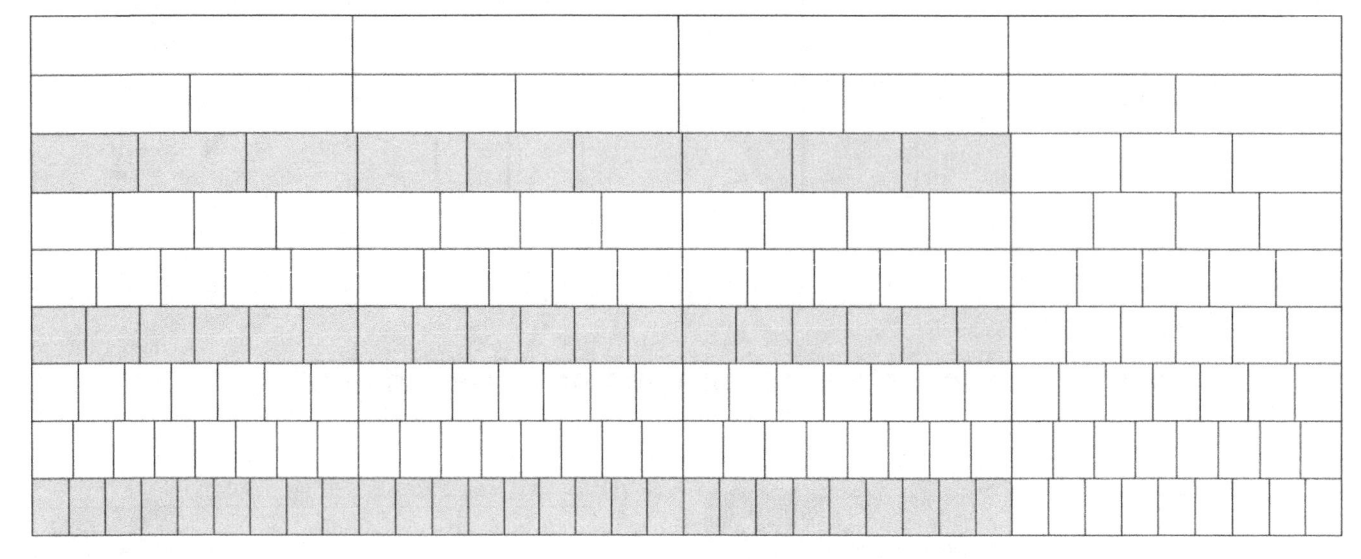

3 strips out of 4 strips are the same as 6 strips out of 8 strips, or 9 strips out of 12 strips, or 12 strips out of 16 strips, or 15 strips out of 20 strips, or 18 strips out of 24 strips, or 21 strips out of 28 strips, or 24 strips out of 32 strips, or 27 strips out of 36 strips - their value remains the same

$$\frac{1}{6} = \frac{2}{12} = \frac{3}{18} = \frac{4}{24} = \frac{5}{30} = \frac{6}{36}$$

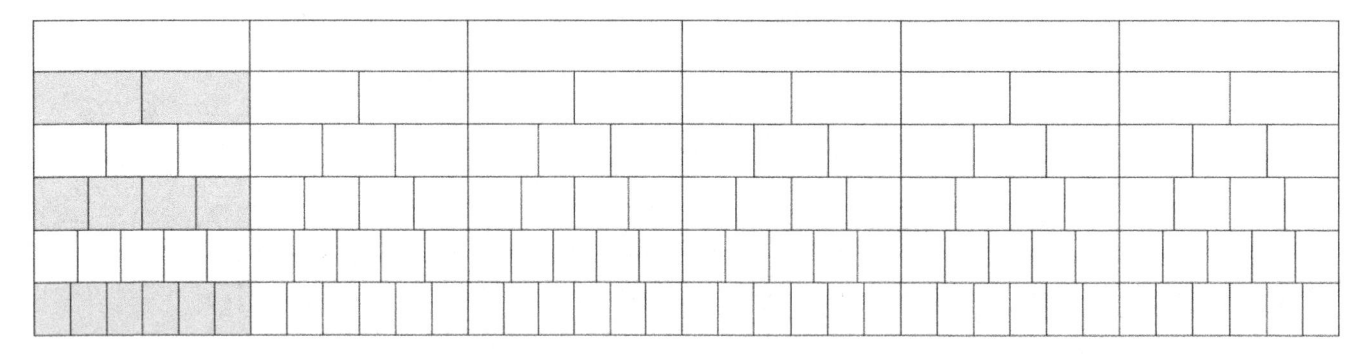

1 strip out of 6 strips is the same as 2 strips out of 12 strips, or 3 strips out of 18 strips, or 4 strips out of 24 strips, or 5 strips out of 30 strips, or 6 strips out of 36 strips - their value remains the same

Aha! The denominators are the same in these fractions! These are common denominators

$$\frac{9}{12} = \quad \frac{18}{24} = \quad \frac{27}{36}$$
$$\frac{2}{12} = \quad \frac{4}{24} = \quad \frac{6}{36}$$

Hint: You have a list of equivalent fractions 😊

Step 1 is done! We changed the given fractions to equivalent fractions to find common denominators. Now we must find the least common denominator.

12 is the LEAST common denominator for

$$\frac{3}{4} \, (= \frac{9}{12}) \text{ and } \frac{1}{6} \, (= \frac{2}{12}).$$

Hint: The least common denominator is the denominator which is common to both denominators of the given fractions.

These are like fractions since they both have 12 in the denominator. So, you can write $\boxed{12}$ as a common denominator for both fractions.

And then, subtract the numerators 9 and 2.

$$\frac{3}{4} - \frac{1}{6} = \frac{9}{12} - \frac{2}{12} = \frac{9-2}{12} = \frac{7}{12}$$

Subtract fractions:

$$\frac{1}{3} - \frac{1}{5} = ?$$

I'm thrilled!

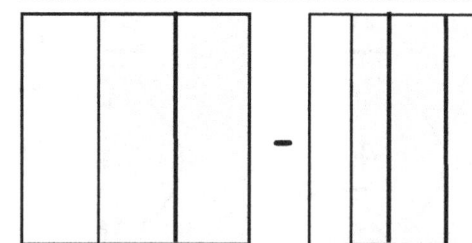

Hint: To subtract fractions, subtract LIKE fractions by simply subtracting the numerators. 😊

Step 1: Find the LEAST common multiple of the denominators.

How to find the least common multiple

For the denominator 3:   1 × 3 = 3   2 × 3 = 6
3 × 3 = 9   4 × 3 = 12   5 × 3 = 15   6 × 3 = 18
For the denominator 5:   1 × 5 = 5   2 × 5 = 10
3 × 5 = 15

Step 1: List all the multiples of the first denominator

Step 2: List all the multiples of the second denominator

The LEAST common multiple of 3 and 5 is 15:   5 × 3 = (15)
3 × 5 = (15)

Step 3: Circle the least multiple which is common to both denominators

Step 2: Draw a fractions line, write 15 for the common denominator, and draw a slash above the numerator of each fraction:

$$\frac{1\backslash}{3} - \frac{1\backslash}{5} = \frac{-}{15} =$$

Hint: To find a common denominator is to raise a fraction in higher terms to find an equivalent fraction (like $\frac{1}{3} - \frac{5}{15} = \frac{?}{15} - \frac{5}{15}$). 😊

Step 3: Divide the least common multiple by each denominator and write the quotient above each slash:

Hint: Think of how many thirds are in fifteenths? How many fifths are in fifteenths?

$$\overset{5}{\diagdown}\frac{1}{3} - \overset{3}{\diagdown}\frac{1}{5} = \frac{\_\_ - \_\_}{15} =$$

$15 \div 3 = 5$

$15 \div 5 = 3$

Step 4: Multiply each numerator by the number above a slash:

$5 \times 1 = 5$

$3 \times 1 = 3$

Hint: You are finding the "new" numerators using a common denominator of 15:

$$\overset{5}{\diagdown}\frac{1}{3} - \overset{3}{\diagdown}\frac{1}{5} = \frac{(5 \times 1) - (3 \times 1)}{15} =$$

You got the "new" numerators".

These are the numerators for the denominator 15:

$$\frac{1}{3} = \frac{5}{15}, \frac{1}{5} = \frac{3}{15}.$$

Step 5: Subtract the "new" numerators":

$$\overset{5}{\diagdown}\frac{1}{3} - \overset{3}{\diagdown}\frac{1}{5} = \frac{5 - 3}{15} = \frac{2}{15}$$

1. <u>Subtract</u> fractions. <u>Write</u> the missing numbers. <u>Color</u> the fraction of each shape. <u>Simplify</u> if possible.

$\dfrac{3}{4} - \dfrac{1}{3} = ?$ $\qquad\qquad\qquad\qquad$ $\dfrac{3}{4} - \dfrac{1}{5} = ?$

Step 1: Find the least common denominator

$\dfrac{3}{4} = \dfrac{6}{\underline{}} = \dfrac{9}{\underline{}}$ $\qquad\qquad$ $\dfrac{3}{4} = \dfrac{6}{\underline{}} = \dfrac{9}{\underline{}} = \dfrac{12}{\underline{}} = \dfrac{15}{\underline{}}$

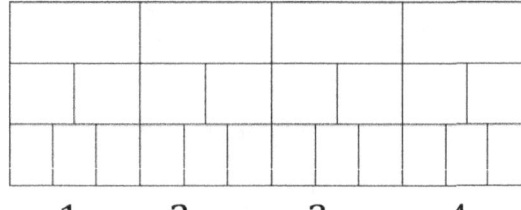

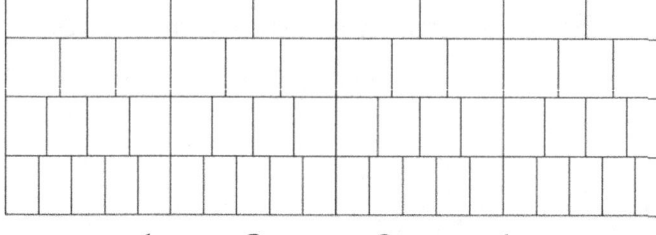

$\dfrac{1}{3} = \dfrac{2}{\underline{}} = \dfrac{3}{\underline{}} = \dfrac{4}{\underline{}}$

$\dfrac{1}{5} = \dfrac{2}{\underline{}} = \dfrac{3}{\underline{}} = \dfrac{4}{\underline{}}$

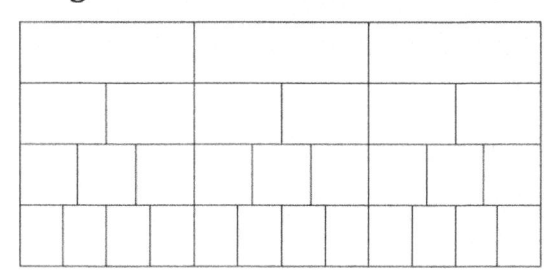

____ is the least common denominator

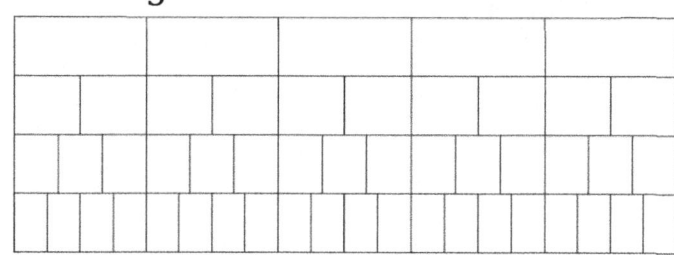

____ is the least common denominator

Step 2: Add the Numerators to find the sum

$\dfrac{3}{4} - \dfrac{1}{3} = \dfrac{-}{\underline{}} = \dfrac{}{\underline{}} = \dfrac{}{\underline{}}$ $\qquad$ $\dfrac{3}{4} - \dfrac{2}{5} = \dfrac{-}{\underline{}} = \dfrac{}{\underline{}} = \dfrac{}{\underline{}}$

2. I had $\boxed{54}$ points in the beginning of the game. I used $\boxed{\dfrac{3}{6}}$ of the points to buy a boat and $\boxed{\dfrac{2}{9}}$ of the leftover points to buy a helicopter. <u>How many points did I use?</u>

Answer: _____

1. <u>Subtract</u> fractions. <u>Write</u> the missing numbers. <u>Color</u> the fraction of each shape. <u>Simplify</u> if possible.

$\frac{2}{3} - \frac{7}{12} = ?$ $\qquad\qquad\qquad\qquad$ $\frac{1}{4} - \frac{1}{6} = ?$

Step 1: Find the least common denominator

$\frac{2}{3} = \frac{4}{\underline{}} = \frac{6}{\underline{}} = \frac{8}{\underline{}}$ $\qquad\qquad$ $\frac{1}{4} = \frac{2}{\underline{}} = \frac{3}{\underline{}}$

$\frac{7}{12}$ $\qquad\qquad\qquad\qquad\qquad\qquad$ $\frac{1}{6} = \frac{2}{\underline{}}$

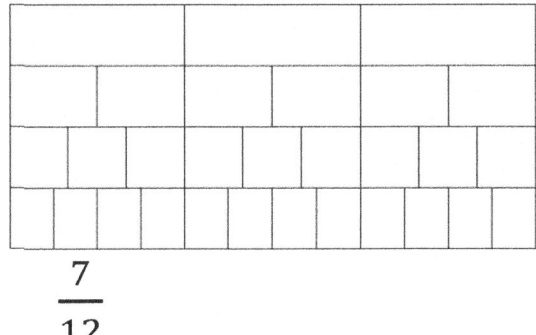

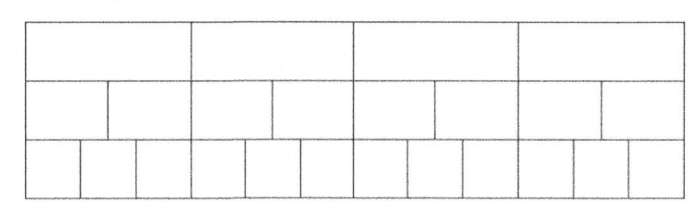

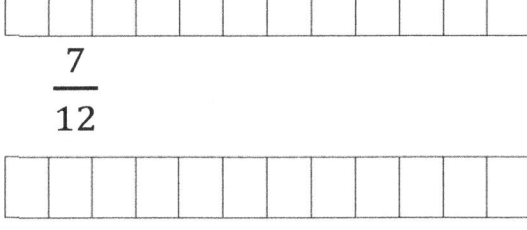

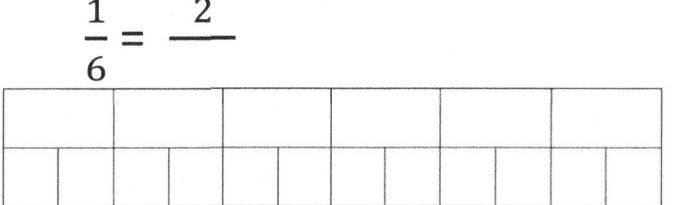

____ is the least common denominator $\qquad$ ____ is the least common denominator

Step 2: Add the Numerators to find the sum

$\frac{2}{3} - \frac{7}{12} = \frac{ - }{\underline{}} = \frac{}{} = \underline{}\,\frac{}{}$ $\qquad$ $\frac{1}{4} - \frac{1}{6} = \frac{ - }{\underline{}} = \frac{}{} = \underline{}\,\frac{}{}$

2. I encountered $\boxed{48}$ monsters in the game. I destroyed $\boxed{\frac{1}{3}}$ of the monsters with my sword and $\boxed{\frac{5}{8}}$ of the leftover monsters with my gun. <u>How many monsters</u> did I destroy?

Answer: _____

1. Subtract fractions. Write the missing numbers. Simplify if possible.

$\dfrac{3}{8} - \dfrac{1}{6} = \dfrac{\quad - \quad}{\quad} = \dfrac{\quad}{\quad} = \_\dfrac{\quad}{\quad}$ $\dfrac{7}{9} - \dfrac{1}{3} = \dfrac{\quad - \quad}{\quad} = \dfrac{\quad}{\quad} = \_\dfrac{\quad}{\quad}$

$\dfrac{5}{8} - \dfrac{1}{4} = \dfrac{\quad - \quad}{\quad} = \dfrac{\quad}{\quad} = \_\dfrac{\quad}{\quad}$ $\dfrac{11}{15} - \dfrac{2}{5} = \dfrac{\quad - \quad}{\quad} = \dfrac{\quad}{\quad} = \_\dfrac{\quad}{\quad}$

$\dfrac{3}{10} - \dfrac{1}{4} = \dfrac{\quad - \quad}{\quad} = \dfrac{\quad}{\quad} = \_\dfrac{\quad}{\quad}$ $\dfrac{11}{16} - \dfrac{3}{8} = \dfrac{\quad - \quad}{\quad} = \dfrac{\quad}{\quad} = \_\dfrac{\quad}{\quad}$

$\dfrac{8}{12} - \dfrac{2}{6} = \dfrac{\quad - \quad}{\quad} = \dfrac{\quad}{\quad} = \_\dfrac{\quad}{\quad}$ $\dfrac{7}{8} - \dfrac{3}{6} = \dfrac{\quad - \quad}{\quad} = \dfrac{\quad}{\quad} = \_\dfrac{\quad}{\quad}$

$\dfrac{13}{18} - \dfrac{2}{3} = \dfrac{\quad - \quad}{\quad} = \dfrac{\quad}{\quad} = \_\dfrac{\quad}{\quad}$ $\dfrac{7}{15} - \dfrac{1}{3} = \dfrac{\quad - \quad}{\quad} = \dfrac{\quad}{\quad} = \_\dfrac{\quad}{\quad}$

$\dfrac{5}{7} - \dfrac{9}{14} = \dfrac{\quad - \quad}{\quad} = \dfrac{\quad}{\quad} = \_\dfrac{\quad}{\quad}$ $\dfrac{9}{10} - \dfrac{5}{6} = \dfrac{\quad - \quad}{\quad} = \dfrac{\quad}{\quad} = \_\dfrac{\quad}{\quad}$

2. Mother used $\boxed{\dfrac{1}{3}}$ of the water bucket to water the roses and $\boxed{\dfrac{7}{24}}$ of the water bucket to water the daisies, and $\boxed{\dfrac{5}{16}}$ of the water bucket to water the tulips. How much water (—) was left?

Answer: _____

1. Write in the missing numbers (add or subtract fractions).

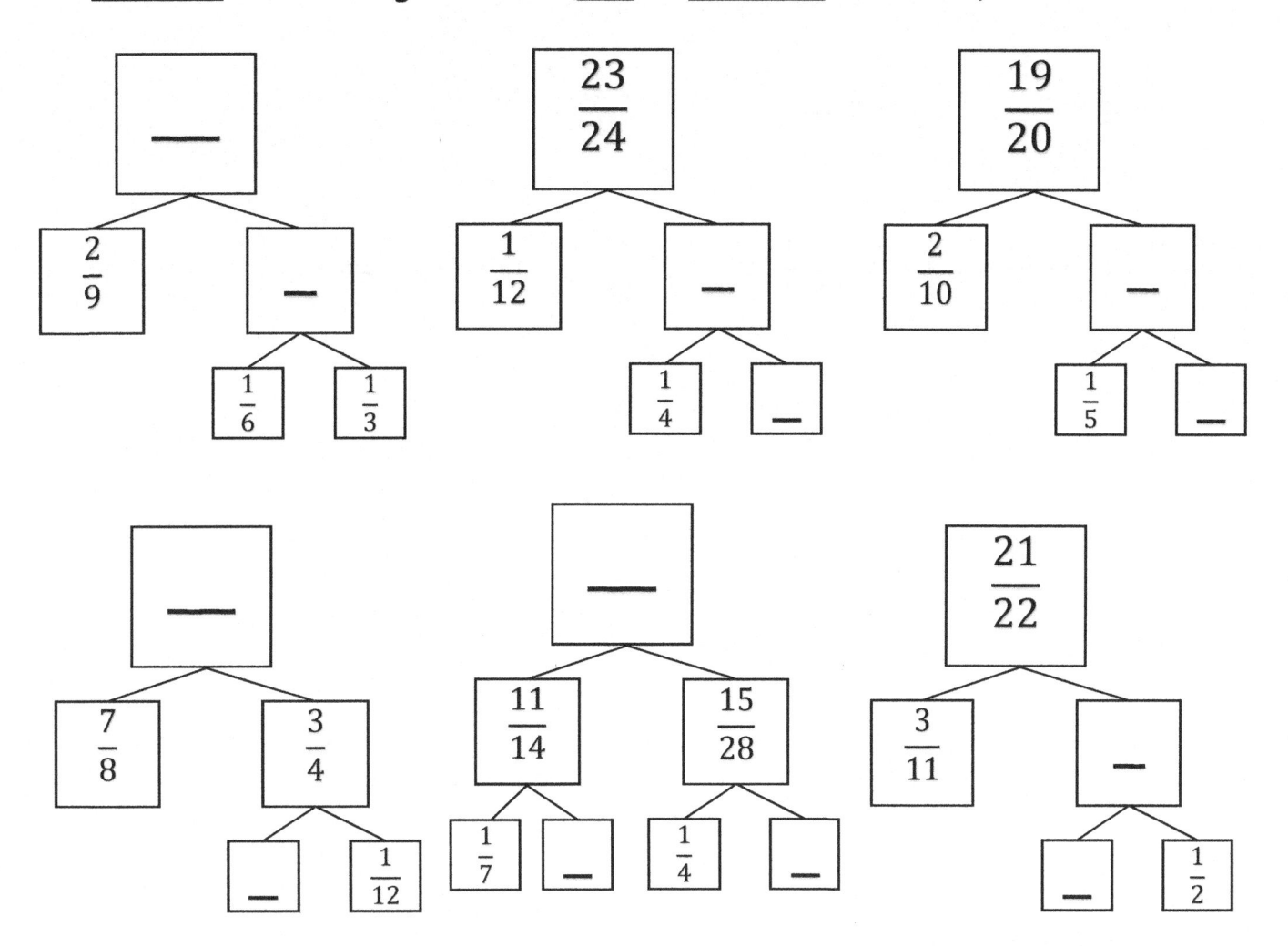

[Fraction tree 1: Top box (empty) → $\frac{2}{9}$ and (empty box) → $\frac{1}{6}$ and $\frac{1}{3}$]

[Fraction tree 2: Top $\frac{23}{24}$ → $\frac{1}{12}$ and (empty) → $\frac{1}{4}$ and (empty)]

[Fraction tree 3: Top $\frac{19}{20}$ → $\frac{2}{10}$ and (empty) → $\frac{1}{5}$ and (empty)]

[Fraction tree 4: Top (empty) → $\frac{7}{8}$ and $\frac{3}{4}$ → (empty) and $\frac{1}{12}$]

[Fraction tree 5: Top (empty) → $\frac{11}{14}$ and $\frac{15}{28}$ → children: $\frac{1}{7}$ and (empty); $\frac{1}{4}$ and (empty)]

[Fraction tree 6: Top $\frac{21}{22}$ → $\frac{3}{11}$ and (empty) → (empty) and $\frac{1}{2}$]

2. I spent $\frac{3}{7}$ of my allowance on a book and $\frac{5}{14}$ of it on a bag. I spent _____ of my allowance altogether. Circle the correct answer.

    a) $\frac{10}{14}$ B) $\frac{9}{14}$ c) $\frac{14}{21}$ d) $\frac{11}{14}$

1. <u>Subtract</u> fractions. <u>Write</u> the missing numbers. <u>Simplify</u> if possible.

$\dfrac{1}{2} - \dfrac{1}{6} = \dfrac{-}{} = \dfrac{-}{-} = \dfrac{}{}$ $\dfrac{13}{30} - \dfrac{2}{5} = \dfrac{-}{} = \dfrac{-}{-} = \dfrac{}{}$

$\dfrac{19}{24} - \dfrac{5}{8} = \dfrac{-}{} = \dfrac{-}{-} = \dfrac{}{}$ $\dfrac{15}{28} - \dfrac{2}{4} = \dfrac{-}{} = \dfrac{-}{-} = \dfrac{}{}$

$\dfrac{3}{4} - \dfrac{3}{12} = \dfrac{-}{} = \dfrac{-}{-} = \dfrac{}{}$ $\dfrac{5}{7} - \dfrac{4}{9} = \dfrac{-}{} = \dfrac{-}{-} = \dfrac{}{}$

$\dfrac{4}{11} - \dfrac{2}{22} = \dfrac{-}{} = \dfrac{-}{-} = \dfrac{}{}$ $\dfrac{7}{10} - \dfrac{3}{40} = \dfrac{-}{} = \dfrac{-}{-} = \dfrac{}{}$

$\dfrac{15}{20} - \dfrac{2}{5} = \dfrac{-}{} = \dfrac{-}{-} = \dfrac{}{}$ $\dfrac{5}{6} - \dfrac{3}{8} = \dfrac{-}{} = \dfrac{-}{-} = \dfrac{}{}$

$\dfrac{10}{15} - \dfrac{6}{12} = \dfrac{-}{} = \dfrac{-}{-} = \dfrac{}{}$ $\dfrac{12}{36} - \dfrac{8}{48} = \dfrac{-}{} = \dfrac{-}{-} = \dfrac{}{}$

2. Peter had some cookies. He ate $\dfrac{2}{9}$ of the cookies and he still had 21 cookies left. <u>How many cookies</u> did Peter have in the beginning?

Answer: _____

1. <u>Subtract</u> fractions. <u>Write</u> the missing numbers. <u>Simplify</u> if possible.

$\dfrac{5}{6} - \dfrac{3}{24} =$ \qquad $\dfrac{7}{9} - \dfrac{1}{3} =$

$\dfrac{5}{9} - \dfrac{1}{27} =$ \qquad $\dfrac{7}{16} - \dfrac{5}{48} =$

$\dfrac{23}{24} - \dfrac{1}{3} - \dfrac{1}{4} = \dfrac{\quad - \quad - \quad}{\quad\quad\quad} = \dfrac{\quad}{\quad}$

$\dfrac{5}{6} - \dfrac{1}{5} - \dfrac{1}{15} = \dfrac{\quad - \quad - \quad}{\quad\quad\quad} = \dfrac{\quad}{\quad}$

$\dfrac{6}{7} - \dfrac{2}{5} - \dfrac{5}{70} = \dfrac{\quad - \quad - \quad}{\quad\quad\quad} = \dfrac{\quad}{\quad}$

$\dfrac{11}{12} - \dfrac{5}{48} - \dfrac{7}{24} = \dfrac{\quad - \quad - \quad}{\quad\quad\quad} = \dfrac{\quad}{\quad}$

2. Kelly had some cards. She traded $\dfrac{3}{8}$ of the cards and she still had 35 cards left. <u>How many cards</u> did Kelly have in the beginning?

Answer: _____

1. Write in the missing numbers (add or subtract).

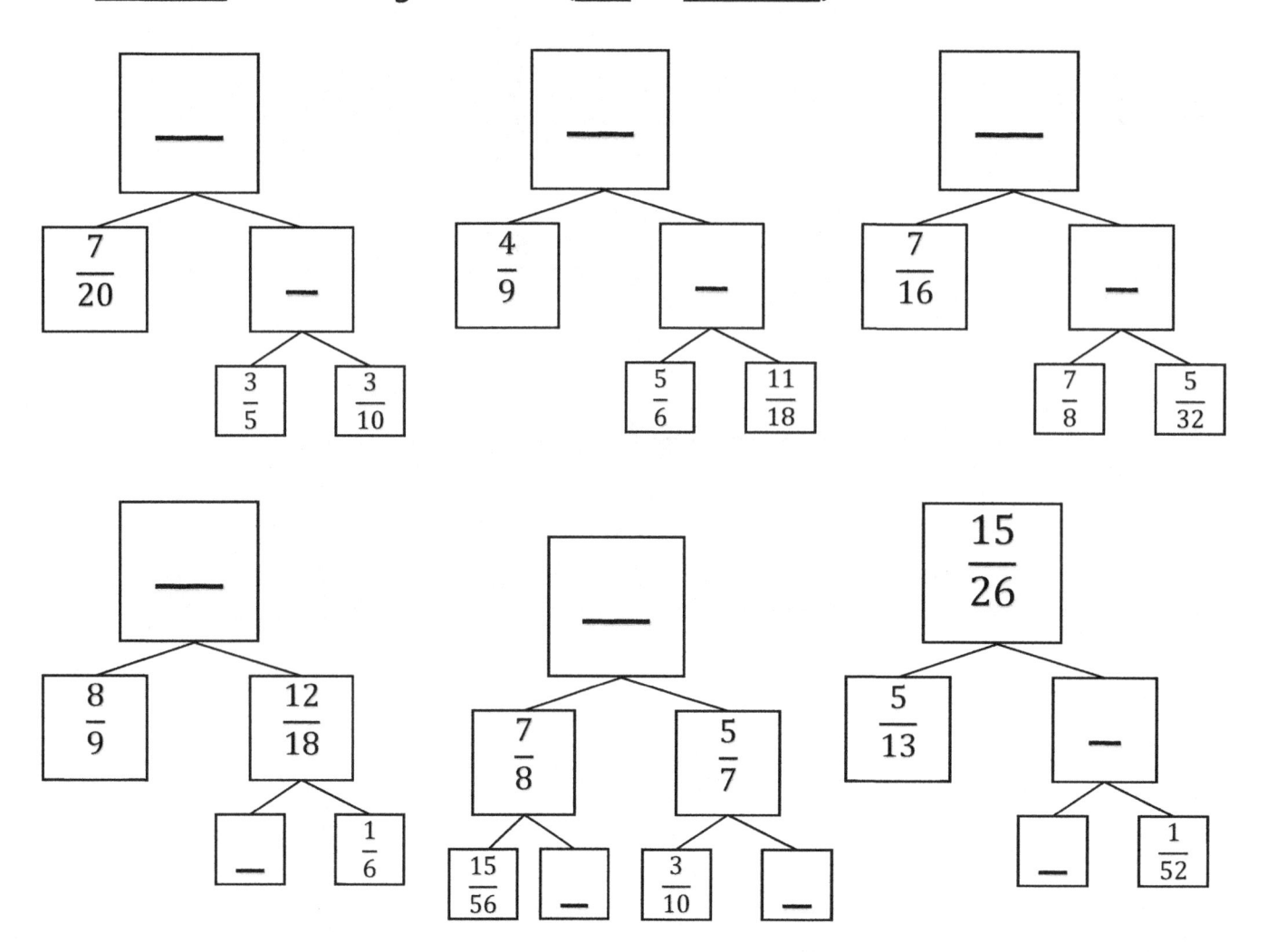

Tree 1: Top = __ ; children: 7/20 and __ ; the __ child has children 3/5 and 3/10.

Tree 2: Top = __ ; children: 4/9 and __ ; the __ child has children 5/6 and 11/18.

Tree 3: Top = __ ; children: 7/16 and __ ; the __ child has children 7/8 and 5/32.

Tree 4: Top = __ ; children: 8/9 and 12/18 ; the 12/18 has children __ and 1/6.

Tree 5: Top = __ ; children: 7/8 and 5/7 ; the 7/8 has children 15/56 and __ ; the 5/7 has children 3/10 and __.

Tree 6: Top = 15/26 ; children: 5/13 and __ ; the __ child has children __ and 1/52.

2. The length of a rectangle is $\frac{1}{2}$ inch shorter than its width. If its width is 3 inches, find its area. Circle the correct answer.

a) 7 b) $\frac{16}{2}$ c) $6\frac{1}{2}$ d) $7\frac{1}{2}$

96

1. <u>Subtract</u> fractions. <u>Write</u> the missing numbers. <u>Simplify</u> if possible.

$1 - \frac{1}{6} = ---- = \frac{-}{} = -$ $1 - \frac{1}{3} = ---- = \frac{-}{} = -$

Hint: To subtract a fraction from a whole number: change a whole number to a fraction with the same denominator as that in the fraction to be subtracted

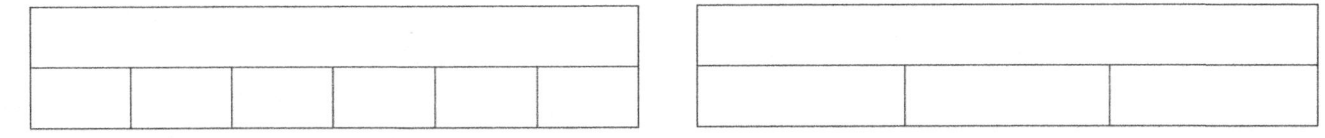

$1 - \frac{3}{4} = ---- = \frac{-}{} = -$ $1 - \frac{2}{5} = ---- = \frac{-}{} = -$

Hint: Rewrite a whole number to a fraction so that both fractions have a common denominator:

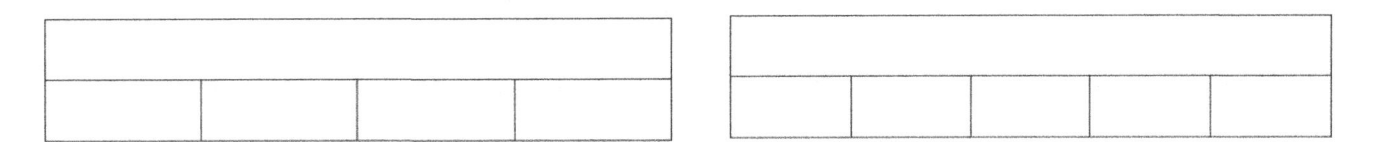

$1 - \frac{5}{7} = ---- = \frac{-}{} = -$ $1 - \frac{3}{8} = ---- = \frac{-}{} = -$

Hint: First, simplify if possible! Then, subtract!

$2 - \frac{2}{6} = \_--- = \frac{-}{} = - = \_$ $3 - \frac{3}{6} = \_--- = \frac{-}{} = - = \_$

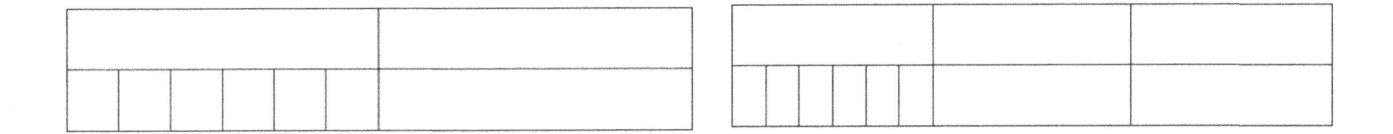

$4 - \frac{8}{12} = \_--- = \frac{-}{} = - = \_$ $5 - \frac{9}{15} = \_--- = \frac{-}{} = - = \_$

1. <u>Subtract</u> fractions. <u>Write</u> the missing numbers. <u>Simplify</u> if possible.

Hint: To subtract a fraction from a whole number: rewrite a whole number to a mixed number: a whole number and a fraction with the same denominator (like $2\frac{9}{9}$).

$3 - \frac{4}{9} = 2\frac{9}{9} - \underline{} = \frac{ - }{} = \underline{}\frac{}{}$ $6 - \frac{8}{12} = \underline{} - \underline{} = \frac{ - }{} = \underline{}\frac{}{}$

$4 - \frac{2}{3} = \underline{} - \underline{} = \frac{ - }{} = \underline{}\frac{}{}$ $5 - \frac{2}{5} = \underline{} - \underline{} = \frac{ - }{} = \underline{}\frac{}{}$

$6 - \frac{9}{14} = \underline{} - \underline{} = \frac{ - }{} = \underline{}\frac{}{}$ $7 - \frac{5}{6} = \underline{} - \underline{} = \frac{ - }{} = \underline{}\frac{}{}$

2. <u>Multiply</u> and <u>simplify</u> if possible. <u>Circle</u> the correct answer (letter).

<u>Find</u> $\frac{10}{18}$ of 54.

$\frac{10}{18} \times 54 = \frac{ \times }{ \times } = \underline{}$

a)   50
b)   44
c)   30

<u>Find</u> $\frac{2}{3}$ of 72.

Step 1: ___ ÷ ___ = ___

Step 2: ___ × ___ = ___

a)   48
b)   52
c)   60

<u>Find</u> $\frac{3}{4}$ of 56.

Step 1: ___ ÷ ___ = ___

Step 2: ___ × ___ = ___

d)   32
e)   40
f)   42

Add mixed numbers: $2\frac{1}{3} + 1\frac{1}{4} = ?$

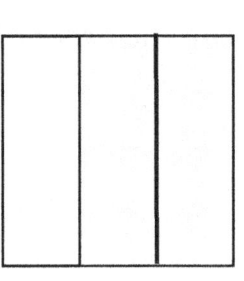

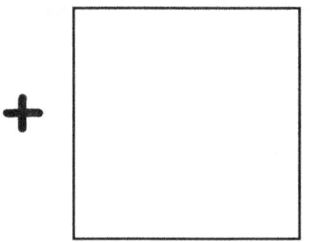

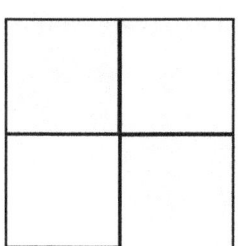

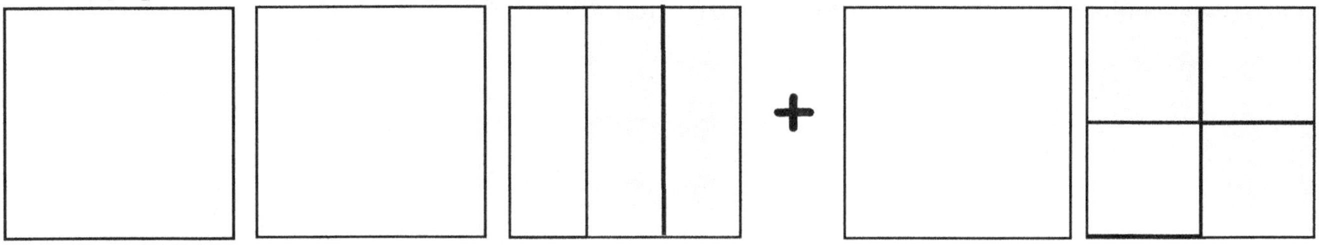

Step 1: Add the whole numbers:

$2 + 1 = 3$.

$2 + 1 = 3$

Hint: To add mixed numbers, make sure that the fractional parts have the common denominator!

Step 2: Change to equivalent fractions finding the LEAST common multiple which is equal to common denominator

For the denominator 3:	$1 \times 3 = 3$	$2 \times 3 = 6$	$3 \times 3 = 9$
	$4 \times 3 = 12$	$5 \times 3 = 15$	$6 \times 3 = 18$
For the denominator 4:	$1 \times 4 = 4$	$2 \times 4 = 8$	$3 \times 4 = 12$

The LEAST common multiple of 3 and 4 is 12:

$4 \times 3 = 12$

$3 \times 4 = 12$

You can say 12 is the least common denominator!

Step 3: Draw a fractions bar, write 12 for the denominator, and draw a slash above the numerator for each fraction:

$\dfrac{1}{3} + \dfrac{1}{4} = \dfrac{\_ + \_}{12} =$

Step 4: Divide the least common multiple by each denominator and write the quotient above each slash:

Hint: Use slashes in the beginning to make the process easy and fun 😊

$$\overset{4}{\diagdown}\frac{1}{3} + \overset{3}{\diagdown}\frac{1}{4} = \frac{\_ + \_}{12} =$$

$$12 \div 3 = 4$$
$$12 \div 4 = 3$$

Step 5: Multiply each numerator by the number above a slash:

$$4 \times 1 = 4$$
$$3 \times 1 = 3$$

Hint: You can also say that you're finding equivalent fractions using a common denominator of 12 😊

$$\overset{4}{\diagdown}\frac{1}{3} + \overset{3}{\diagdown}\frac{1}{4} = \frac{(4 \times 1) + (3 \times 1)}{12} =$$

You got the "new" numerators".

These are the numerators for the denominator of 12:

$$\frac{1}{3} = \frac{4}{12}; \frac{1}{4} = \frac{3}{12}.$$

Step 6: Add the "new" numerators":

$$\overset{4}{\diagdown}\frac{1}{3} + \overset{3}{\diagdown}\frac{1}{4} = \frac{4 + 3}{12} = \frac{7}{12}$$

Step 7: Add the whole number and a fractional part:

$$3 + \frac{7}{12} = 3\frac{7}{12}$$

1. Add fractions. Write the missing numbers. Simplify if possible.

$2\frac{3}{8} + 4\frac{5}{9} = \_\frac{\_}{\_}$

   1) Add whole numbers   2) Add fractions

1)   2)

$3\frac{3}{4} + 5\frac{3}{5} = \_\frac{\_}{\_}$

1)   2)

$6\frac{3}{10} + 4\frac{5}{6} = \_\frac{\_}{\_}$

1)   2)

2. Change the mixed number to an improper faction. Circle the correct answer.

$3\frac{7}{13} = \frac{\_}{\_}$ $8\frac{5}{9} = \frac{\_}{\_}$

a) $\frac{39}{13}$ b) $\frac{23}{13}$ c) $\frac{46}{13}$ a) $\frac{72}{9}$ b) $\frac{75}{9}$ c) $\frac{77}{9}$

Subtract mixed numbers: $2\frac{1}{3} - 1\frac{2}{8} = ?$

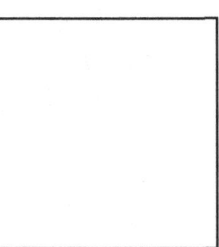

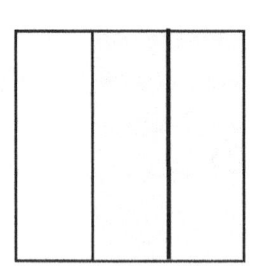

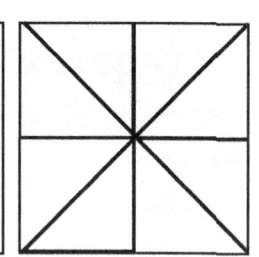

Step 1: Subtract the whole numbers: 2 - 1 = 1.

$2 - 1 = 1$

Hint: To subtract mixed numbers, make sure that the fractional parts have the common denominator!

Step 2: Change to equivalent fractions finding the LEAST common multiple which is equal to common denominator.

For the denominator 3:	$1 \times 3 = 3$	$2 \times 3 = 6$	$3 \times 3 = 9$
	$4 \times 3 = 12$	$5 \times 3 = 15$	$6 \times 3 = 18$
	$7 \times 3 = 21$	$8 \times 3 = 24$	
For the denominator 4:	$1 \times 8 = 8$	$2 \times 8 = 16$	$3 \times 8 = 24$

The LEAST common multiple of 3 and 8 is 24:

$8 \times 3 = 24$

$3 \times 8 = 24$

You can say 24 is the least common denominator!

Step 3: Draw a fractions bar, write 12 for the denominator, and draw a slash above the numerator for each fraction:

$\dfrac{1}{3} - \dfrac{2}{8} = \dfrac{\_\_ - \_\_}{24} =$

Step 4: Divide the least common multiple by each denominator and write the quotient above each slash:

Hint: Use slashes in the beginning to make the process easy and fun 😊

$$\frac{1}{3}^{\diagdown 8} - \frac{2}{8}^{\diagdown 3} = \frac{\_\_ - \_\_}{24} =$$

$$24 \div 3 = 8$$
$$24 \div 8 = 3$$

Step 5: Multiply each numerator by the number above a slash:

$$8 \times 1 = 8$$
$$3 \times 2 = 6$$

Your "new" numerators

Hint: You can also say that you're finding the equivalent fractions using a common denominator of 24 😊

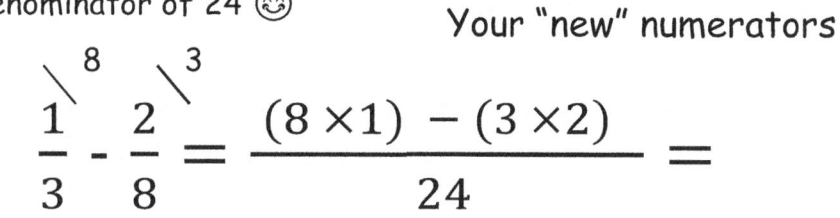

$$\frac{1}{3}^{\diagdown 8} - \frac{2}{8}^{\diagdown 3} = \frac{(8 \times 1) - (3 \times 2)}{24} =$$

You got the "new" numerators".

These are the numerators for the denominator 12:

$$\frac{1}{3} = \frac{8}{24}, \frac{2}{8} = \frac{6}{24}.$$

Step 6: Add the "new" numerators" and simplify:

$$\frac{1}{3}^{\diagdown 8} - \frac{2}{8}^{\diagdown 3} = \frac{8 - 6}{24} = \frac{2}{24} = \frac{1}{12}$$

Step 7: Add the whole number and a fractional part:

$$1 + \frac{1}{12} = 1\frac{1}{12}$$

1. <u>Subtract</u> fractions. <u>Write</u> the missing numbers. <u>Simplify</u> if possible.

$3\frac{5}{8} - 1\frac{5}{6} = \_\frac{\_}{\_}$

1) <u>Subtract</u> whole numbers   2) <u>Subtract</u> fractions

To subtract a greater fraction from a smaller fraction: Step 1: borrow one from a whole number; Step 2: rewrite the borrowed whole number to a fraction: $3\frac{1}{3} - 1\frac{3}{8} = 1\frac{23}{24}$

Step 1: $3-1=\cancel{2}\ 1$ Step 2: $\frac{1}{3} - \frac{3}{8} = \frac{8-9}{24} = \frac{8+24-9}{24} = \frac{23}{24}$

1)   2)

$8\frac{7}{10} - 2\frac{3}{4} = \_\frac{\_}{\_}$

1)   2)

$9\frac{1}{15} - 3\frac{7}{9} = \_\frac{\_}{\_}$

1)   2)

2. <u>Change</u> the improper faction to a mixed number. <u>Circle</u> the correct answer.

$\frac{32}{5} = \_\frac{\_}{\_}$ $\frac{49}{11} = \_\frac{\_}{\_}$

a) $3\frac{2}{5}$ b) $6\frac{1}{5}$ c) $6\frac{2}{5}$ a) $11\frac{5}{11}$ b) $1\frac{38}{11}$ c) $4\frac{5}{11}$

1. <u>Multiply</u> fractions. <u>Write</u> the missing numbers. <u>Simplify</u> if possible.

Step 1: <u>Change</u> the mixed number to an improper fraction

Step 2: <u>Multiply</u> the numerators and denominators

$$3 \times 2\frac{3}{15} = \frac{3}{1} \times \frac{33}{15} = \frac{3}{1} \times \frac{11}{5} = \frac{11+11+11}{5} = \_\frac{\_}{\_}$$

Hint: Simplify before multiplying where you can! (cancel the 3s from 33 and 15)

Hint: Change the mixed number to improper fractions!

$$6 \times 3\frac{5}{12} = \frac{\_}{\_} \times \frac{\_}{\_} = \frac{\_ \times \_}{\_ \times \_} = \frac{\_}{\_} = \_\frac{\_}{\_}$$

Hint: Cancel or cross out the common factors of the numerator and the denominator to simplify where it's possible 😊

$$5 \times 2\frac{7}{20} = \frac{\_}{\_} \times \frac{\_}{\_} = \frac{\_ \times \_}{\_ \times \_} = \frac{\_}{\_} = \_\frac{\_}{\_}$$

$$4 \times 5\frac{5}{16} = \frac{\_}{\_} \times \frac{\_}{\_} = \frac{\_ \times \_}{\_ \times \_} = \frac{\_}{\_} = \_\frac{\_}{\_}$$

2. <u>Multiply</u>. <u>Express</u> your answer in simplest form.

$$\frac{9}{16} \times \frac{20}{36}$$

a) $\frac{15}{36}$ b) $\frac{5}{16}$ c) $\frac{180}{432}$ d) $2\frac{2}{5}$

Math Success Practice Workbook Fractions 4th 5th 6th Grades

1. <u>Multiply</u> fractions. <u>Write</u> the missing numbers. <u>Simplify</u> if possible.

Step 1: <u>Change</u> the mixed number to an improper fraction

Step 2: <u>Multiply</u> the numerators and denominators

$$4\frac{1}{5} \times 4\frac{2}{7} = \frac{}{} \times \frac{}{} = \frac{ \times }{ \times } = \frac{}{} = \underline{}\frac{}{}$$

Hint: Factor the numerator and the denominator by the same number to reduce the fraction to lowest terms! ☺

$$2\frac{1}{4} \times 3\frac{4}{6} = \frac{}{} \times \frac{}{} = \frac{ \times }{ \times } = \frac{}{} = \underline{}\frac{}{}$$

$$3\frac{1}{2} \times 2\frac{2}{7} = \frac{}{} \times \frac{}{} = \frac{ \times }{ \times } = \frac{}{} = \underline{}\frac{}{}$$

$$9\frac{5}{8} \times 3\frac{7}{11} = \frac{}{} \times \frac{}{} = \frac{ \times }{ \times } = \frac{}{} = \underline{}\frac{}{}$$

2. <u>Multiply</u>. <u>Express</u> your answer in simplest form.

$$\frac{5}{12} \times \frac{36}{55}$$

a) $\frac{3}{10}$ b) $\frac{36}{60}$ c) $\frac{3}{11}$ d) $1\frac{2}{5}$

© 2020 Home Run Press, LLC

1. <u>Write in</u> the missing numbers (<u>practice</u> adding and subtracting mixed numbers).

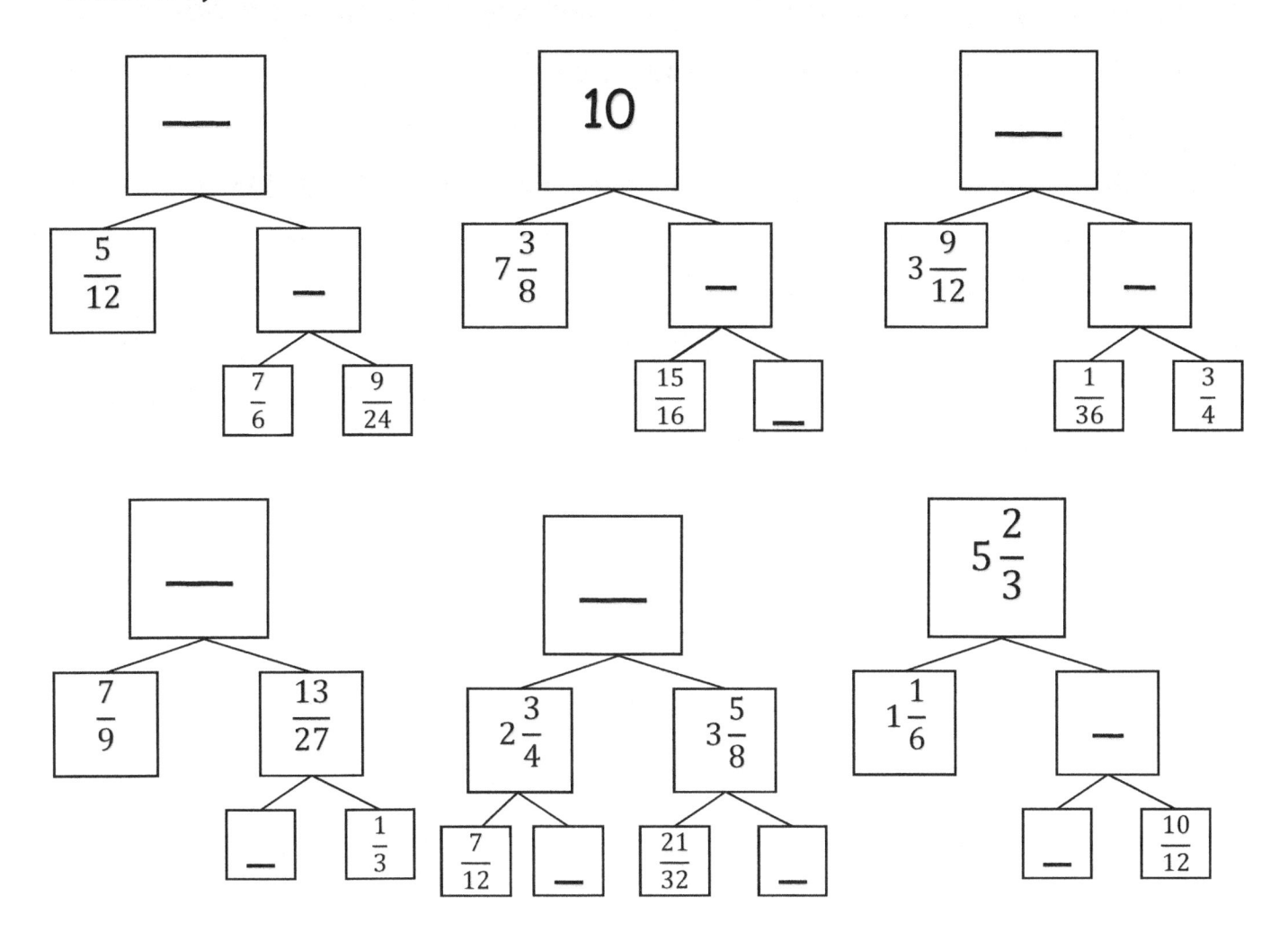

Tree 1: Top = ___; children: $\frac{5}{12}$ and ___; second child's children: $\frac{7}{6}$ and $\frac{9}{24}$.

Tree 2: Top = 10; children: $7\frac{3}{8}$ and ___; second child's children: $\frac{15}{16}$ and ___.

Tree 3: Top = ___; children: $3\frac{9}{12}$ and ___; second child's children: $\frac{1}{36}$ and $\frac{3}{4}$.

Tree 4: Top = ___; children: $\frac{7}{9}$ and $\frac{13}{27}$; second child's children: ___ and $\frac{1}{3}$.

Tree 5: Top = ___; children: $2\frac{3}{4}$ and $3\frac{5}{8}$; first child's children: $\frac{7}{12}$ and ___; second child's children: $\frac{21}{32}$ and ___.

Tree 6: Top = $5\frac{2}{3}$; children: $1\frac{1}{6}$ and ___; second child's children: ___ and $\frac{10}{12}$.

2. A class is $\frac{5}{8}$ filled with spectators. $\frac{2}{5}$ of them are boys and the rest are girls. The fraction of girls in the class is _____.

a) $1\frac{1}{40}$ B) $\frac{3}{5}$ c) $\frac{9}{40}$ d) $\frac{3}{8}$

© 2020 Home Run Press, LLC

1. Divide fractions. Write the missing numbers. Simplify if possible.

Divide fractions! That's impossible! I cannot divide 3 blocks by, say, $\frac{5}{9}$!

$3 \div \frac{5}{9}$ = no answer, no solution, let's skip it.

Your problem means: "How many $\frac{5}{9}$'s are there in 3?" or "How many $\frac{5}{9}$'s are contained in 3?", or "How many $\frac{5}{9}$'s in 3?".

Step 1: Change 3 to $\frac{1}{9}$'s. Use a common denominator of 9. $3 = \frac{27}{9}$

Step 2: Divide the numerators, then, divide the denominators, then, divide the numerator by the denominator.

So, $3 \div \frac{5}{9} = \frac{27}{9} \div \frac{5}{9} = \frac{(27 \div 5)}{(9 \div 9)} = \frac{(27 \div 5)}{1} = \frac{27}{5} = 5\frac{2}{5}$

```
              5 = the whole number
          5 | 27 = Numerator
Denominator
             25
             ___
              2 = Remainder
```

Dividend = Numerator
Divisor = Denominator
Quotient = The Whole Number
Remainder = Numerator of a Fraction Part

$5 \div \frac{2}{7} = \frac{35}{7} \div \frac{2}{7} = \frac{(\quad \div \quad)}{(\quad \div \quad)} = \frac{}{} = \underline{}\,\frac{}{}$

$4 \div \frac{2}{3} = \frac{}{} \div \frac{}{} = \frac{(\quad \div \quad)}{(\quad \div \quad)} = \frac{}{} = \underline{}\,\frac{}{}$

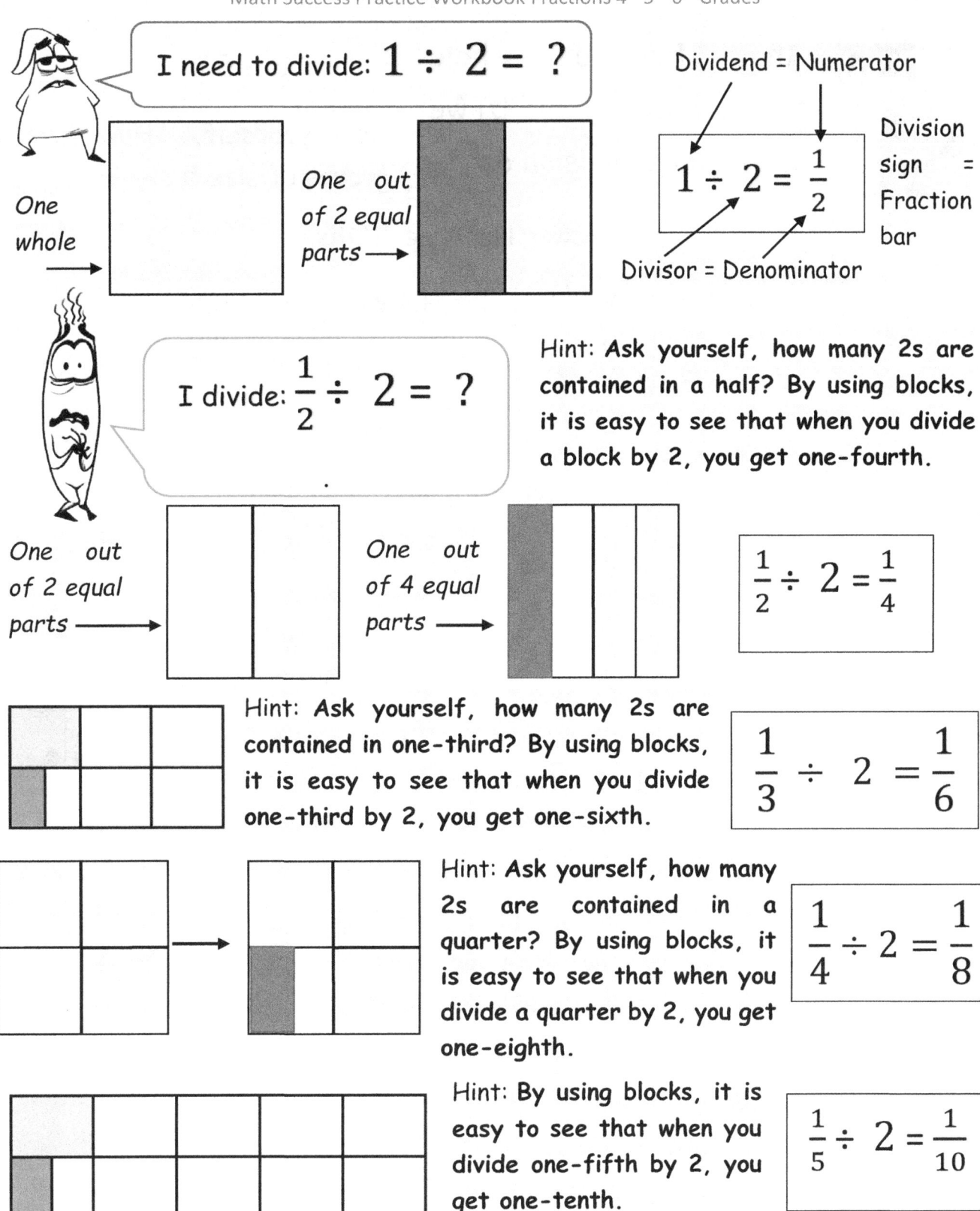

I need to divide: $1 \div 2 = ?$

One whole

One out of 2 equal parts →

Dividend = Numerator
Divisor = Denominator

$1 \div 2 = \dfrac{1}{2}$

Division sign = Fraction bar

I divide: $\dfrac{1}{2} \div 2 = ?$

One out of 2 equal parts →

One out of 4 equal parts →

$\dfrac{1}{2} \div 2 = \dfrac{1}{4}$

Hint: Ask yourself, how many 2s are contained in a half? By using blocks, it is easy to see that when you divide a block by 2, you get one-fourth.

Hint: Ask yourself, how many 2s are contained in one-third? By using blocks, it is easy to see that when you divide one-third by 2, you get one-sixth.

$\dfrac{1}{3} \div 2 = \dfrac{1}{6}$

Hint: Ask yourself, how many 2s are contained in a quarter? By using blocks, it is easy to see that when you divide a quarter by 2, you get one-eighth.

$\dfrac{1}{4} \div 2 = \dfrac{1}{8}$

Hint: By using blocks, it is easy to see that when you divide one-fifth by 2, you get one-tenth.

$\dfrac{1}{5} \div 2 = \dfrac{1}{10}$

Hint: By using blocks, it is easy to see that when you divide one-fifth by 2, you get one-tenth.

$\dfrac{1}{6} \div 2 = \dfrac{1}{12}$

Aha... For all whole numbers (like 3 and 5) we can rewrite division as:

$$3 \div 5 = \frac{3}{5}$$

Ask yourself, "How many groups of size 5 are in 3?"

The answer is: three fifths.

I need to divide:

$$\frac{3}{4} \div 3 = ?$$

How many 3s are contained in $\frac{3}{4}$?

By using blocks, it is easy to see that when you divide three quarters by 3, you get one quarter.

$$\frac{3}{4} \div 3 = \frac{1}{4}$$

I need to divide:

$$\frac{1}{2} \div 4 = ?$$

How many 4s are contained in $\frac{1}{2}$?

By using blocks, it is easy to see that when you divide a half by 4, you get one-eighth.

$$\frac{1}{2} \div 4 = \frac{1}{8}$$

1. How many 8s are contained in $\frac{2}{4}$?

Write the missing numbers. Fill in the fractions in the table.

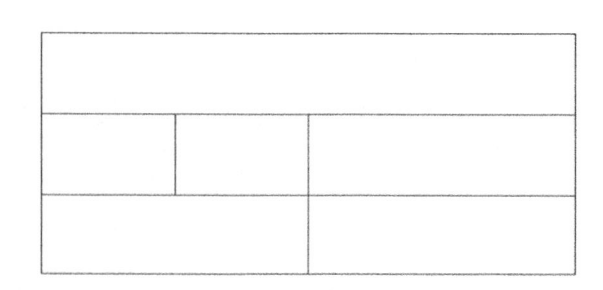

$$\frac{2}{4} \div 8 = \text{—}$$

Can you show me some rules to do division problems easier?

Turn a fraction upside down. It is often called taking the RECIPROCAL. Reciprocal is a number (fraction) where you interchange the numerator and the denominator.

Write the missing numbers to get a reciprocal.

$$\frac{1}{4} \to \frac{4}{1} \qquad \frac{3}{8} \to — \qquad \frac{5}{9} \to — \qquad \frac{6}{11} \to —$$

Hint: To find a reciprocal you need to invert the fraction so that the numerator and the denominator switched their places. 😊

Find the reciprocal for 1: $\qquad 1 = \frac{1}{1} \to —$

Write the missing numbers to get a reciprocal.   First, find an improper fraction, then, a reciprocal:

$$12 \to \frac{1}{12} \qquad \frac{7}{23} \to — \qquad 9\frac{7}{8} \to —$$

Hint: If you multiply a number by its reciprocal, you get 1:

$$\frac{4}{9} \times \frac{9}{4} = \frac{\cancel{4} \times \cancel{9}}{\cancel{9} \times \cancel{4}} = \frac{1 \times 1}{1 \times 1} = 1$$

I draw the arrows to help myself.

Hint: Cancel 9s from 9 and 9. Cancel 4s from 4 and 4.

Multiply the given fractions and invert the divisor!

$$\frac{5}{8} \div 10 = \frac{5}{8} \div \frac{10}{1} = \frac{5}{8} \times \frac{1}{10} = \frac{\cancel{5} \times 1}{8 \times \cancel{10}^2} = \frac{1}{16}$$

Math Success Practice Workbook Fractions 4th 5th 6th Grades

1. <u>Divide</u> fractions. <u>Write</u> the missing numbers. <u>Simplify</u> if possible.

$$\frac{10}{3} \div 2 = \frac{10}{3} \div \frac{2}{1} = \frac{10}{3} \times \frac{1}{2} = \frac{\_ \times \_}{\_ \times \_} = \frac{\_}{\_}$$

$$\frac{8}{15} \div 4 = \frac{\_}{\_} \div \frac{\_}{\_} = \frac{\_}{\_} \times \frac{\_}{\_} = \frac{\_ \times \_}{\_ \times \_} = \frac{\_}{\_}$$

$$\frac{2}{3} \div 4 = \frac{\_}{\_} \div \frac{\_}{\_} = \frac{\_}{\_} \times \frac{\_}{\_} = \frac{\_ \times \_}{\_ \times \_} = \frac{\_}{\_}$$

$$\frac{5}{6} \div 10 = \qquad\qquad\qquad \frac{13}{18} \div 26 =$$

2. The perimeter of a rectangle is $1\frac{5}{12}$ in². <u>Find</u> the length of the rectangle if its width is $\frac{3}{8}$ in long.

Answer: _____

112 © 2020 Home Run Press, LLC

1. <u>Divide</u> fractions. <u>Write</u> the missing numbers. <u>Simplify</u> if possible.

$\dfrac{49}{50} \div 7 = \dfrac{}{} \div \dfrac{}{} = \dfrac{}{} \times \dfrac{}{} = \dfrac{ \times }{ \times } = \dfrac{}{}$

$\dfrac{14}{18} \div 2 = \dfrac{}{} \div \dfrac{}{} = \dfrac{}{} \times \dfrac{}{} = \dfrac{ \times }{ \times } = \dfrac{}{}$

$\dfrac{8}{27} \div 18 =$ $\dfrac{6}{14} \div 21 =$

$\dfrac{21}{25} \div 14 =$ $\dfrac{12}{13} \div 24 =$

2. I had originally $30 and spent $\dfrac{7}{15}$ of it at the café and $\dfrac{1}{5}$ of it at the movies. <u>How much money</u> did I have left?

Answer: _____

1. <u>Divide</u> fractions. <u>Write</u> the missing numbers. <u>Simplify</u> if possible.

$$\frac{9}{17} \div 18 = \frac{}{} \div \frac{}{} = \frac{}{} \times \frac{}{} = \frac{ \times }{ \times } = \frac{}{}$$

$$\frac{24}{35} \div 3 = \frac{}{} \div \frac{}{} = \frac{}{} \times \frac{}{} = \frac{ \times }{ \times } = \frac{}{}$$

$$\frac{49}{50} \div x = \frac{7}{50} \qquad\qquad \frac{15}{16} \div x = \frac{3}{64}$$

$$x = \frac{49}{50} \div \frac{7}{50} = \frac{49}{50} \times \frac{50}{7} = \frac{}{} \qquad\qquad x =$$

$$\frac{36}{41} \div x = \frac{4}{41} \qquad\qquad \frac{25}{50} \div x = \frac{1}{30}$$

$$x = \qquad\qquad\qquad\qquad x =$$

2. A jar $\frac{5}{6}$ full contains 2 L of water. How much water will the jar contain when $\frac{3}{5}$ full?

Answer: _____

I need to divide: $1 \div \dfrac{1}{2} = ?$

Think: How many halves are there in one whole?

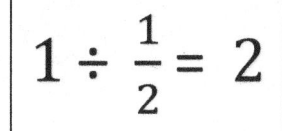

One whole →

Two halves in a block →

$1 \div \dfrac{1}{2} = 2$

$1 \div \dfrac{1}{2} = \dfrac{1}{1} \div \dfrac{1}{2} = \dfrac{1}{1} \times \dfrac{2}{1} = \dfrac{2}{1}$

I divide: $2 \div \dfrac{1}{2} = ?$

Think: How many halves are there in two wholes?

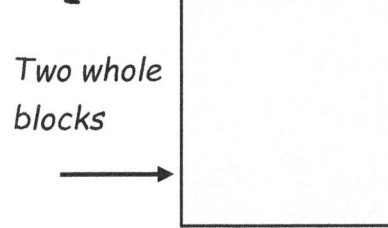

There are 2 halves in 1 whole. Therefore, there must be 4 halves in 2 wholes.

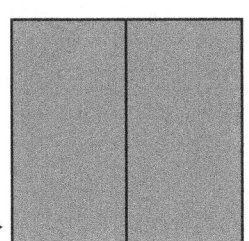

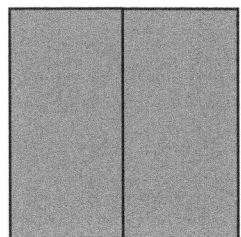

Two whole blocks →

Four halves in the two blocks →

$2 \div \dfrac{1}{2} = 4$

$2 \div \dfrac{1}{2} = \dfrac{2}{1} \div \dfrac{1}{2} = \dfrac{2}{1} \times \dfrac{2}{1} = \dfrac{4}{1}$

$4 \div \dfrac{1}{3} = ?$

Think: How many thirds are there in 4?

To divide a whole number by a fraction:

1) change a whole number so that both fractions have a common denominator

2) or turn the second fraction upside down (= use the reciprocal).

$4 \div \dfrac{1}{3} = \dfrac{12}{3} \div \dfrac{1}{3} = \dfrac{(12 \div 1)}{(3 \div 3)} = \dfrac{12}{1}$

$4 \div \dfrac{1}{3} = \dfrac{4}{1} \div \dfrac{1}{3} = \dfrac{4}{1} \times \dfrac{3}{1} = \dfrac{12}{1}$

1	1	1	1
$\frac{1}{3}$ $\frac{1}{3}$ $\frac{1}{3}$	$\frac{1}{3}$ $\frac{1}{3}$ $\frac{1}{3}$	$\frac{1}{3}$ $\frac{1}{3}$ $\frac{1}{3}$	$\frac{1}{3}$ $\frac{1}{3}$ $\frac{1}{3}$

© 2020 Home Run Press, LLC

Math Success Practice Workbook Fractions 4th 5th 6th Grades

1. <u>Divide</u> fractions. <u>Write</u> the missing numbers. <u>Simplify</u> if possible.

$8 \div \dfrac{2}{7} = \dfrac{8}{1} \div \dfrac{2}{7} = \dfrac{8}{1} \times \dfrac{7}{2} =$ ___-___

$6 \div \dfrac{12}{13} = \dfrac{}{} \div \dfrac{}{} = \dfrac{}{} \times \dfrac{}{} =$ ___-___

$5 \div \dfrac{10}{17} = \dfrac{}{} \div \dfrac{}{} = \dfrac{}{} \times \dfrac{}{} =$ ___-___

$4 \div \dfrac{12}{21} = \dfrac{}{} \div \dfrac{}{} = \dfrac{}{} \times \dfrac{}{} =$ ___-___

$9 \div \dfrac{3}{8} = \dfrac{}{} \div \dfrac{}{} = \dfrac{}{} \times \dfrac{}{} =$ ___-___

2. I had x points and spent $\dfrac{3}{4}$ of them on comic books. <u>How many points</u> did I have if I spent 12 points?

$\dfrac{3}{4} x = 12$

$x = 12 \div \dfrac{3}{4} = \dfrac{}{} \div \dfrac{}{} = \dfrac{}{} \times \dfrac{}{} = \dfrac{ \times }{ \times } =$ ___-___

Answer: _____

1. <u>Divide</u> fractions. <u>Write</u> the missing numbers. <u>Simplify</u> if possible.

$3 \div \frac{21}{25} = \frac{}{} \div \frac{}{} = \frac{}{} \times \frac{}{} = \frac{}{}$

$7 \div \frac{14}{17} = \frac{}{} \div \frac{}{} = \frac{}{} \times \frac{}{} = \frac{}{}$

$3 \div \frac{6}{11} = \frac{}{} \div \frac{}{} = \frac{}{} \times \frac{}{} = \frac{}{}$

$6 \div \frac{24}{25} = \qquad\qquad\qquad 9 \div \frac{18}{22} =$

$3 \div \frac{15}{18} = \qquad\qquad\qquad 10 \div \frac{10}{15} =$

2. Grandma planted x flowers and $\frac{3}{5}$ of them were beautiful 30 lilacs. <u>How many flowers</u> did she plant?

$\frac{3}{5} x = 30$

$x = \underline{} \div \frac{}{} = \frac{}{} \div \frac{}{} = \frac{}{} \times \frac{}{} = \frac{}{}$

Answer: _____

Math Success Practice Workbook Fractions 4ᵗʰ 5ᵗʰ 6ᵗʰ Grades

1. <u>Divide</u> or <u>multiply</u>. <u>Write</u> the missing numbers. <u>Simplify</u> if possible.

$x \div \dfrac{5}{7} = 7$ $\qquad\qquad\qquad\qquad$ $x \div \dfrac{4}{15} = 22\dfrac{1}{2}$

Hint: Change the mixed number to an improper fraction, then divide or multiply!

$x = \dfrac{}{} \times \dfrac{}{} = \dfrac{ \times }{ \times } = \dfrac{}{}$ $\qquad$ $x =$

$x \div \dfrac{24}{25} = 8\dfrac{1}{3}$ $\qquad\qquad\qquad$ $x \div \dfrac{18}{22} = 11$

$x =$ $\qquad\qquad\qquad\qquad\qquad\qquad$ $x =$

$x \div \dfrac{24}{35} = 8\dfrac{3}{4}$ $\qquad\qquad\qquad$ $x \div \dfrac{15}{36} = 28\dfrac{4}{5}$

$x =$ $\qquad\qquad\qquad\qquad\qquad\qquad$ $x =$

2. I read $\boxed{x}$ books and $\boxed{\dfrac{5}{12}}$ of them were $\boxed{35}$ comic books. <u>How many books</u> did I read?

$\dfrac{}{} x = \underline{}$

$x = \underline{} \div \dfrac{}{} = \dfrac{}{} \div \dfrac{}{} = \dfrac{}{} \times \dfrac{}{} = \dfrac{}{}$

Answer: _____

118

I need to divide: $\dfrac{2}{3} \div \dfrac{1}{6} = ?$

Think: How many sixths are there in two thirds?

$$\dfrac{2}{3} \div \dfrac{1}{6} = \dfrac{2}{\cancel{3}} \times \dfrac{\cancel{6}^2}{1} = \dfrac{4}{1}$$

1				
$\dfrac{2}{3}$				
$\dfrac{1}{6}$	$\dfrac{1}{6}$	$\dfrac{1}{6}$	$\dfrac{1}{6}$	

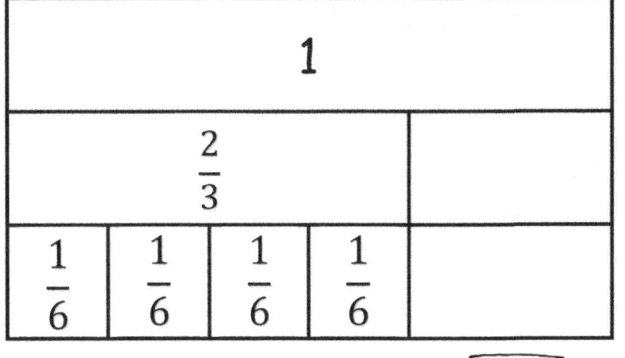

I need to divide $\dfrac{5}{12} \div \dfrac{1}{4} =$

I'll show you the algorithm.

Step 1: Rewrite the first fraction:

$$\dfrac{5}{12} \div \dfrac{1}{4} = \dfrac{5}{12}$$

Step 2: Change the operation (÷ → ×):

$$\dfrac{5}{12} \div \dfrac{1}{4} = \dfrac{5}{12} \times$$

Step 3: Write the reciprocal of the second fraction – switch the numerator and the denominator:

$$\dfrac{5}{12} \div \dfrac{1}{4} = \dfrac{5}{12} \times \dfrac{4}{1}$$

Step 4: Multiply the numerators and denominators and simplify where possible:

$$\dfrac{5}{12} \div \dfrac{1}{4} = \dfrac{5}{\cancel{12}_3} \times \dfrac{4}{1}$$

Step 5: Change to mixed fraction where possible:

$$\dfrac{5}{12} \div \dfrac{1}{4} = \dfrac{5}{\cancel{12}_3} \times \dfrac{4}{1} = 6\dfrac{2}{3}$$

1. Divide fractions. Write the missing numbers. Simplify if possible.

$\dfrac{1}{12} \div \dfrac{5}{6} = \dfrac{}{} \times \dfrac{}{} = \dfrac{}{}$

$\dfrac{8}{9} \div \dfrac{24}{45} = \dfrac{}{} \times \dfrac{}{} = \dfrac{}{}$

$\dfrac{5}{18} \div \dfrac{8}{9} = \dfrac{}{} \times \dfrac{}{} = \dfrac{}{}$

First, change the mixed number to an improper fraction, then, divide or multiply!

$6\dfrac{5}{6} \div 2\dfrac{22}{30} = \dfrac{}{} \div \dfrac{}{} = \dfrac{}{} \times \dfrac{}{} = \dfrac{}{}$

$1\dfrac{8}{9} \div 2\dfrac{14}{27} = $ $4\dfrac{3}{7} \div 4\dfrac{6}{14} = $

2. I spent $\dfrac{1}{3}$ of an hour to find a cave. Then, I spent $\dfrac{5}{6}$ of an hour to dig out 2 diamonds. Then, I came back to my tiny house, found the recipe of a diamond sword, and spent $\dfrac{1}{4}$ of an hour to make it. How much time did I spend in all?

Answer: _____

1. <u>Divide</u> fractions. <u>Write</u> the missing numbers. <u>Simplify</u> if possible.

$$\frac{14}{18} \div \frac{2}{9} = \frac{}{} \times \frac{}{} = \underline{} \frac{}{}$$

$$\frac{6}{7} \div \frac{24}{49} = \qquad\qquad 3\frac{5}{6} \div 4\frac{2}{7} =$$

$$\frac{21}{25} \div \frac{14}{15} = \qquad\qquad 2\frac{3}{8} \div 2\frac{1}{3} =$$

$$1\frac{8}{9} \div 2\frac{2}{33} = \qquad\qquad 4\frac{3}{7} \div 3\frac{6}{14} =$$

$$8\frac{3}{4} \div 3\frac{6}{8} = \qquad\qquad 10\frac{2}{3} \div 5\frac{4}{6} =$$

2. A jar is $\frac{3}{10}$ full of water. If 2 gallons are added, then it is $\frac{4}{5}$ full. <u>How many gallons</u> does the jar hold when full?

Answer: _____

1. <u>Divide</u> fractions. <u>Write</u> the missing numbers. <u>Simplify</u> if possible.

$\frac{16}{25} \div \frac{32}{55} = \frac{}{} \times \frac{}{} = \frac{}{}$

$\frac{6}{7} \div \frac{24}{49} =$ $\qquad$ $5\frac{1}{7} \div 4\frac{1}{2} =$

$\frac{21}{25} \div \frac{14}{15} =$ $\qquad$ $11\frac{3}{7} \div 4\frac{10}{14} =$

$15\frac{1}{8} \div 2\frac{1}{5} =$ $\qquad$ $4\frac{1}{5} \div 2\frac{1}{3} =$

$x \div 3\frac{4}{8} = 2\frac{2}{7}$ $\qquad$ $x \div 7\frac{1}{7} = 3\frac{9}{25}$

$x =$ $\qquad\qquad\qquad\qquad$ $x =$

2. I save $\frac{7}{9}$ of my weekly pocket money. Of the remainder, I spend $\frac{3}{5}$ of it on movies. <u>What fraction of the total money</u> do I spend on the movies?

Answer: _____

Page 8

1. <u>Read and write</u> the missing numbers.

The *numerator* is the numeral that is written above the bar. → $\frac{8}{8}$

$\downarrow \frac{1}{1}$ $\downarrow \frac{2}{2}$ $\frac{4}{4}$ $1=\frac{1}{1}$ $\downarrow \frac{8}{8}$

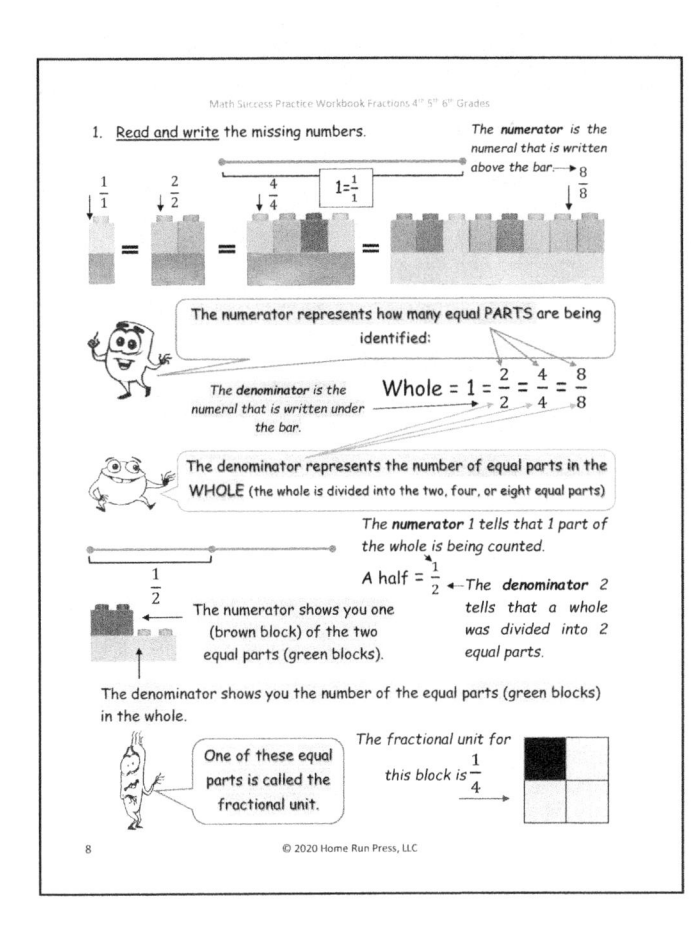

The numerator represents how many equal PARTS are being identified:

The *denominator* is the numeral that is written under the bar.

$$\text{Whole} = 1 = \frac{2}{2} = \frac{4}{4} = \frac{8}{8}$$

The denominator represents the number of equal parts in the WHOLE (the whole is divided into the two, four, or eight equal parts)

The *numerator 1* tells that 1 part of the whole is being counted.

$\frac{1}{2}$

A half = $\frac{1}{2}$ ← The *denominator 2* tells that a whole was divided into 2 equal parts.

The numerator shows you one (brown block) of the two equal parts (green blocks).

The denominator shows you the number of the equal parts (green blocks) in the whole.

One of these equal parts is called the fractional unit.

The fractional unit for this block is $\frac{1}{4}$

Page 9

Aha... I start with multiplying the fractions, right? So, to multiply the fractions I need to multiply out the numerators and then, I need to multiply out the denominators.

Multiply $\frac{2}{3}$ of $\frac{4}{5}$: Add 4 two times: 4 + 4 → $\frac{2 \times 4}{3 \times 5} = \frac{8}{15}$
Add 5 three times: 5 + 5 + 5 →

$\frac{4}{5}$ Four fifths of one whole

$\frac{2}{3}$ Two thirds of one whole

$\frac{2}{3}$ of $\frac{4}{5} = \frac{8}{15}$ Eight fifteenths of one whole

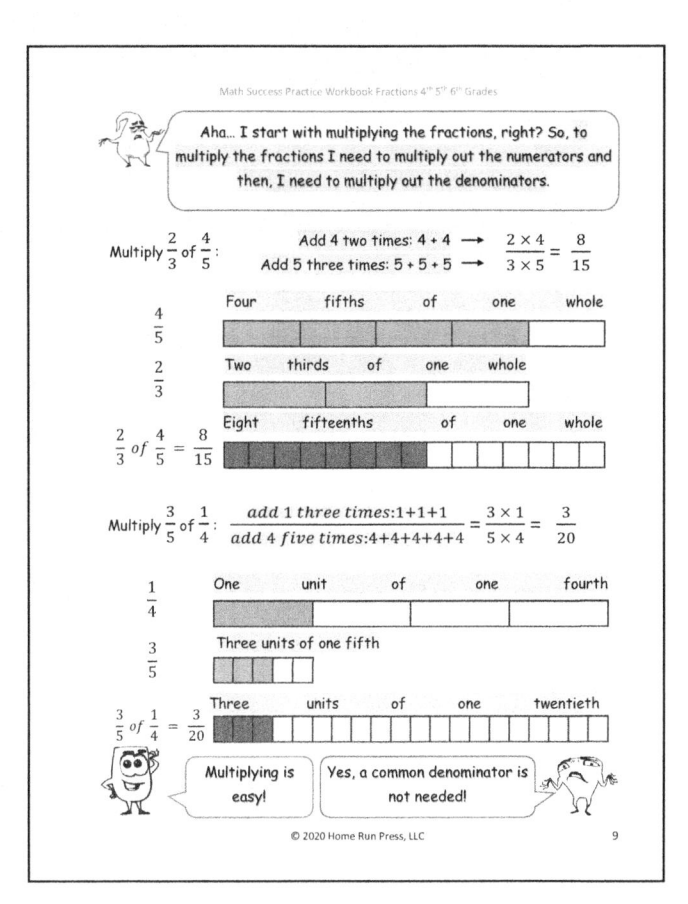

Multiply $\frac{3}{5}$ of $\frac{1}{4}$: $\dfrac{add\ 1\ three\ times: 1+1+1}{add\ 4\ five\ times: 4+4+4+4+4} = \frac{3 \times 1}{5 \times 4} = \frac{3}{20}$

$\frac{1}{4}$ One unit of one fourth

$\frac{3}{5}$ Three units of one fifth

$\frac{3}{5}$ of $\frac{1}{4} = \frac{3}{20}$ Three units of one twentieth

Multiplying is easy!   Yes, a common denominator is not needed!

Page 10

1. <u>Multiply</u> fractions. <u>Color</u> the fractions bars. <u>Write</u> the missing numbers and words.

$\dfrac{2 \times 2}{4 \times 5} = \dfrac{add\ 2\ two\ times: 2+2}{add\ 5\ four\ times: 5+5+5+5} = \frac{4}{20}$

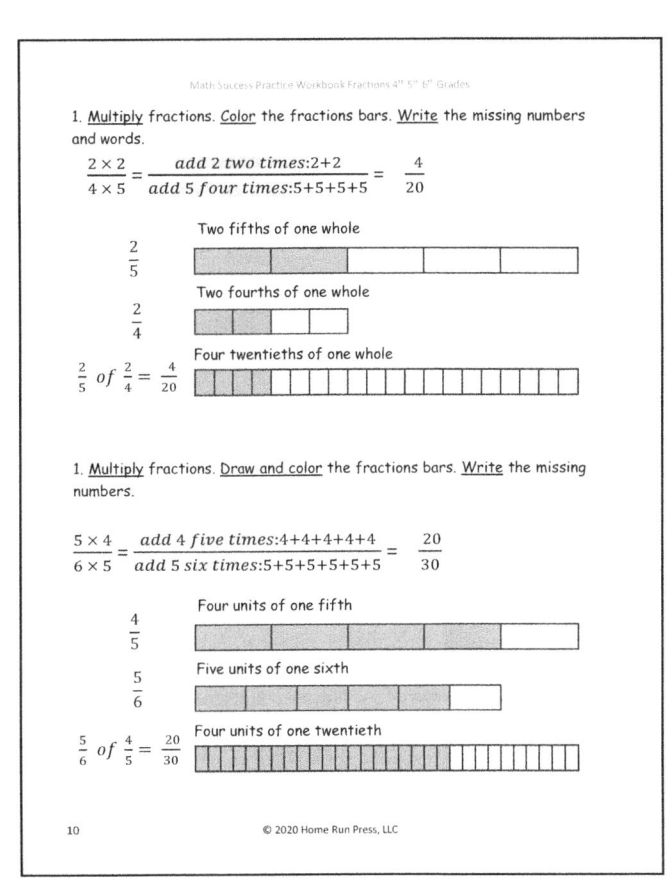

$\frac{2}{5}$ Two fifths of one whole

$\frac{2}{4}$ Two fourths of one whole

$\frac{2}{5}$ of $\frac{2}{4} = \frac{4}{20}$ Four twentieths of one whole

1. <u>Multiply</u> fractions. <u>Draw and color</u> the fractions bars. <u>Write</u> the missing numbers.

$\dfrac{5 \times 4}{6 \times 5} = \dfrac{add\ 4\ five\ times: 4+4+4+4+4}{add\ 5\ six\ times: 5+5+5+5+5+5} = \frac{20}{30}$

$\frac{4}{5}$ Four units of one fifth

$\frac{5}{6}$ Five units of one sixth

$\frac{5}{6}$ of $\frac{4}{5} = \frac{20}{30}$ Four units of one twentieth

Page 11

1. Equivalent fractions. <u>Answer</u> the questions and <u>write</u> the missing numbers.

Equivalent fractions? Oh, boy! What's that?

How can I explain? Equivalent fractions show the same value of the whole.

I circled one third black. The whole is divided into 3 equal parts. $\frac{1}{3}$ $\frac{3}{9}$ I circled three ninths black. The whole is divided into 9 equal parts. I circled 9 dots in both pictures.

The numerators and denominators are different, but the value is the same. Take a half of the block: $\frac{1}{2}$ (a half). If you multiply the numerator and the denominator BY THE SAME NUMBER called the COMMON FACTOR, say, 2, you'll get a different fraction - $\frac{2}{4}$ (two fourths), but their value will stay the same:

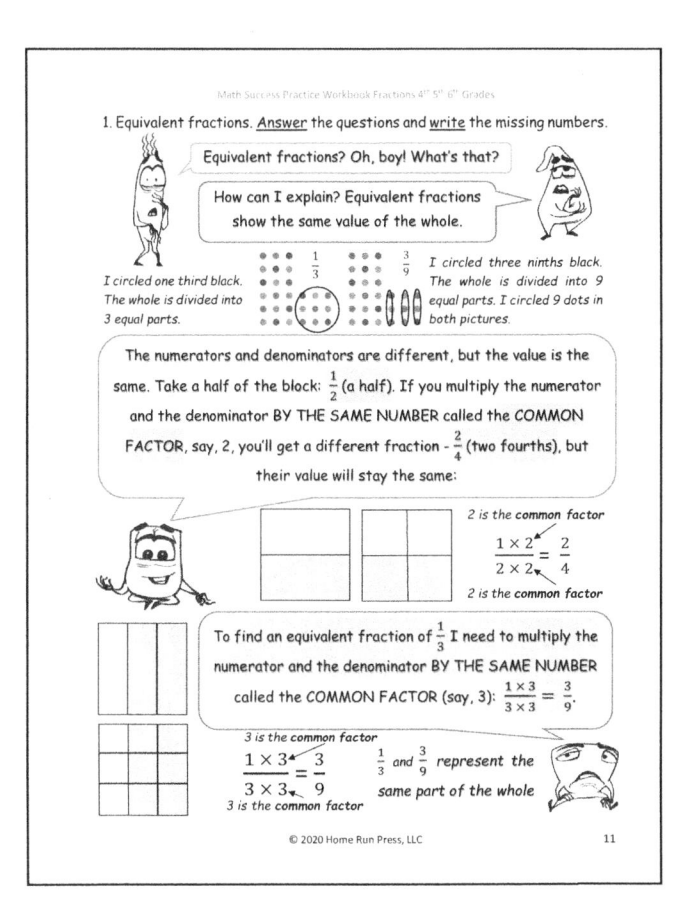

2 is the common factor
$\dfrac{1 \times 2}{2 \times 2} = \dfrac{2}{4}$
2 is the common factor

To find an equivalent fraction of $\frac{1}{3}$ I need to multiply the numerator and the denominator BY THE SAME NUMBER called the COMMON FACTOR (say, 3): $\dfrac{1 \times 3}{3 \times 3} = \dfrac{3}{9}$.

3 is the common factor
$\dfrac{1 \times 3}{3 \times 3} = \dfrac{3}{9}$ $\frac{1}{3}$ and $\frac{3}{9}$ represent the same part of the whole
3 is the common factor

Page 12

1. <u>Fill in</u> the missing numbers using the fractions strips.

$\dfrac{1 \times 2}{2 \times 2} = \dfrac{2}{4}$

$\dfrac{1 \times 4}{2 \times 4} = \dfrac{4}{8}$

$\dfrac{1 \times 2}{3 \times 2} = \dfrac{2}{6}$

$\dfrac{1 \times 2}{5 \times 2} = \dfrac{2}{10}$

$\dfrac{1 \times 2}{4 \times 2} = \dfrac{2}{8}$

1in

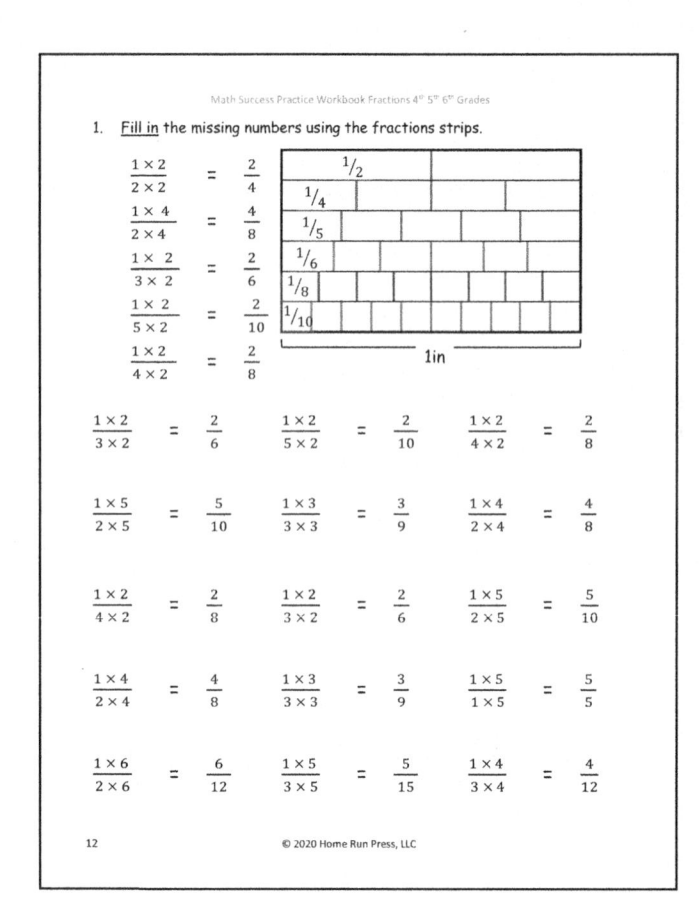

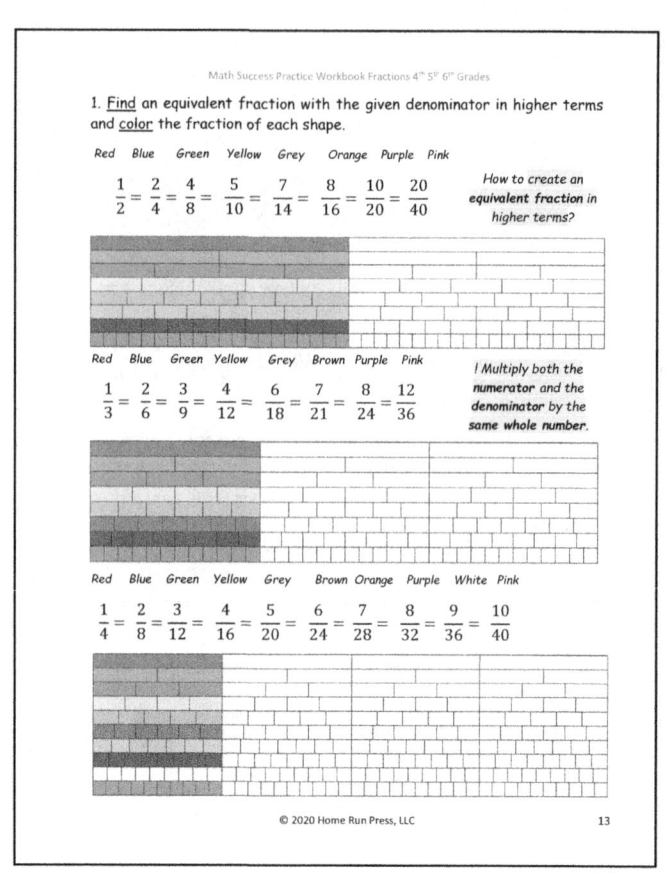

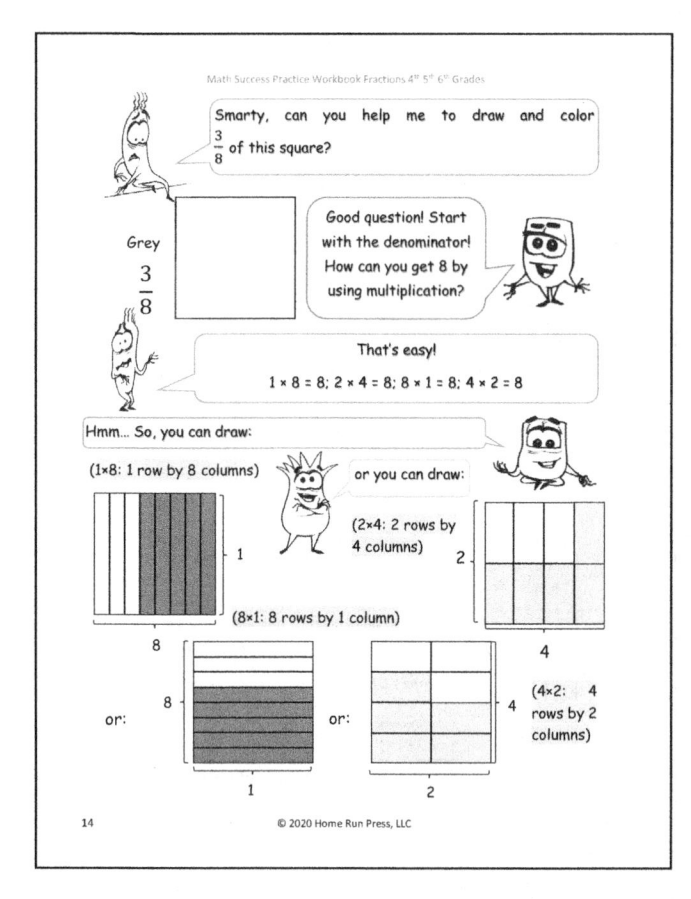

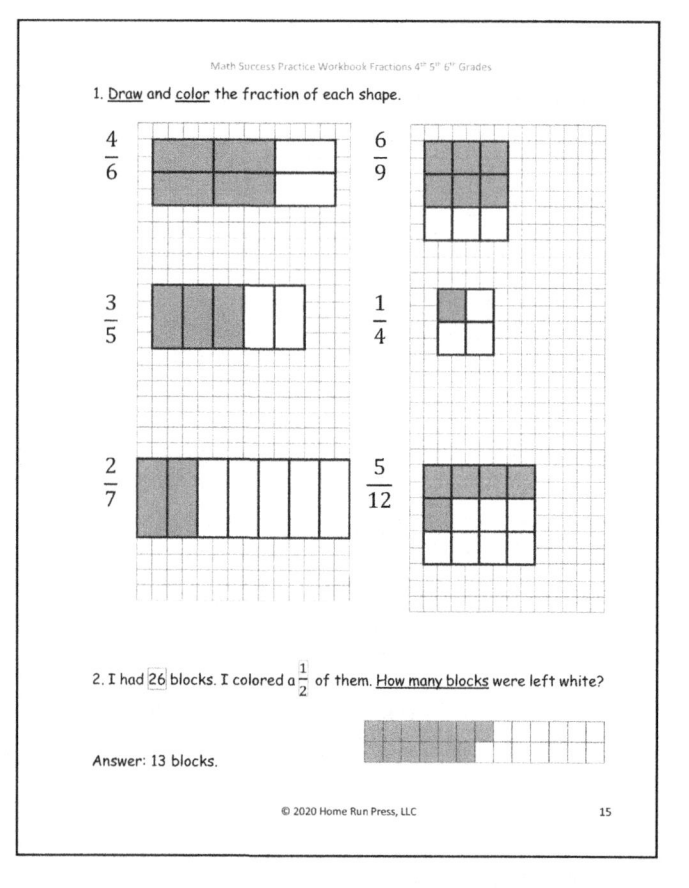

$\dfrac{1 \times 2}{3 \times 2} = \dfrac{2}{6}$ $\dfrac{1 \times 2}{5 \times 2} = \dfrac{2}{10}$ $\dfrac{1 \times 2}{4 \times 2} = \dfrac{2}{8}$

$\dfrac{1 \times 5}{2 \times 5} = \dfrac{5}{10}$ $\dfrac{1 \times 3}{3 \times 3} = \dfrac{3}{9}$ $\dfrac{1 \times 4}{2 \times 4} = \dfrac{4}{8}$

$\dfrac{1 \times 2}{4 \times 2} = \dfrac{2}{8}$ $\dfrac{1 \times 2}{3 \times 2} = \dfrac{2}{6}$ $\dfrac{1 \times 5}{2 \times 5} = \dfrac{5}{10}$

$\dfrac{1 \times 4}{2 \times 4} = \dfrac{4}{8}$ $\dfrac{1 \times 3}{3 \times 3} = \dfrac{3}{9}$ $\dfrac{1 \times 5}{1 \times 5} = \dfrac{5}{5}$

$\dfrac{1 \times 6}{2 \times 6} = \dfrac{6}{12}$ $\dfrac{1 \times 5}{3 \times 5} = \dfrac{5}{15}$ $\dfrac{1 \times 4}{3 \times 4} = \dfrac{4}{12}$

Page 13

1. <u>Find</u> an equivalent fraction with the given denominator in higher terms and <u>color</u> the fraction of each shape.

Red   Blue   Green   Yellow   Grey   Orange   Purple   Pink

$\dfrac{1}{2} = \dfrac{2}{4} = \dfrac{4}{8} = \dfrac{5}{10} = \dfrac{7}{14} = \dfrac{8}{16} = \dfrac{10}{20} = \dfrac{20}{40}$

*How to create an **equivalent fraction** in higher terms?*

Red   Blue   Green   Yellow   Grey   Brown   Purple   Pink

$\dfrac{1}{3} = \dfrac{2}{6} = \dfrac{3}{9} = \dfrac{4}{12} = \dfrac{6}{18} = \dfrac{7}{21} = \dfrac{8}{24} = \dfrac{12}{36}$

*I Multiply both the **numerator** and the **denominator** by the same whole number.*

Red   Blue   Green   Yellow   Grey   Brown   Orange   Purple   White   Pink

$\dfrac{1}{4} = \dfrac{2}{8} = \dfrac{3}{12} = \dfrac{4}{16} = \dfrac{5}{20} = \dfrac{6}{24} = \dfrac{7}{28} = \dfrac{8}{32} = \dfrac{9}{36} = \dfrac{10}{40}$

Page 14

Smarty, can you help me to draw and color $\dfrac{3}{8}$ of this square?

Grey $\dfrac{3}{8}$

Good question! Start with the denominator! How can you get 8 by using multiplication?

That's easy!
$1 \times 8 = 8$; $2 \times 4 = 8$; $8 \times 1 = 8$; $4 \times 2 = 8$

Hmm... So, you can draw:

(1×8: 1 row by 8 columns)

or you can draw:

(2×4: 2 rows by 4 columns)

(8×1: 8 rows by 1 column)

(4×2: 4 rows by 2 columns)

Page 15

1. <u>Draw</u> and <u>color</u> the fraction of each shape.

$\dfrac{4}{6}$ $\dfrac{6}{9}$

$\dfrac{3}{5}$ $\dfrac{1}{4}$

$\dfrac{2}{7}$ $\dfrac{5}{12}$

2. I had <u>26</u> blocks. I colored a $\dfrac{1}{2}$ of them. <u>How many blocks</u> were left white?

Answer: 13 blocks.

Page 16

1. <u>Find</u> an equivalent fraction with the given numerator.

Step 1: Use the given fraction ($\frac{1}{2}$).
Step 2: The numerator 1 has been multiplied by 2 to get 2.
Step 3: To get an equivalent fraction, the denominator 2 must also be multiplied by 2 to get 4.
Therefore $\frac{1}{2} = \frac{2}{4}$.

$\frac{1}{2} = \frac{2}{4} = \frac{11}{22} = \frac{4}{8} = \frac{9}{18} = \frac{6}{12} = \frac{10}{20} = \frac{8}{16} = \frac{5}{10} = \frac{7}{14} = \frac{3}{6} = \frac{12}{24}$

$\frac{1}{3} = \frac{2}{6} = \frac{12}{36} = \frac{4}{12} = \frac{8}{24} = \frac{11}{33} = \frac{7}{21} = \frac{5}{15} = \frac{9}{27} = \frac{10}{30} = \frac{6}{18} = \frac{3}{9}$

$\frac{1}{4} = \frac{9}{36} = \frac{3}{12} = \frac{12}{48} = \frac{5}{20} = \frac{8}{32} = \frac{11}{44} = \frac{6}{24} = \frac{2}{8} = \frac{10}{40} = \frac{7}{28} = \frac{4}{16}$

$\frac{1}{5} = \frac{8}{40} = \frac{3}{15} = \frac{12}{60} = \frac{2}{10} = \frac{10}{50} = \frac{4}{20} = \frac{5}{25} = \frac{9}{45} = \frac{6}{30} = \frac{11}{55} = \frac{7}{35}$

2. <u>Write</u> the fraction name under each picture.

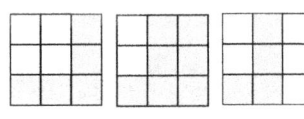

$\frac{white\ blocks}{blocks\ in\ total}$

$\frac{4}{9}$ $\frac{2}{9}$ $\frac{5}{9}$

Page 17

1. <u>Find</u> an equivalent fraction with the given denominator.

Step 1: Use the given fraction ($\frac{1}{2}$).
Step 2: The denominator 2 has been multiplied by 9 to get 18.
Step 3: To get an equivalent fraction, the numerator 1 must also be multiplied by 9 to get 9.
Therefore $\frac{1}{2} = \frac{9}{18}$.

$\frac{1}{2} = \frac{9}{18} = \frac{2}{4} = \frac{6}{12} = \frac{11}{22} = \frac{8}{16} = \frac{4}{8} = \frac{14}{28} = \frac{20}{40} = \frac{3}{6} = \frac{5}{10} = \frac{17}{34}$

$\frac{1}{3} = \frac{7}{21} = \frac{3}{9} = \frac{20}{60} = \frac{5}{15} = \frac{9}{27} = \frac{4}{12} = \frac{2}{6} = \frac{6}{18} = \frac{12}{36} = \frac{16}{48} = \frac{8}{24}$

$\frac{1}{4} = \frac{6}{24} = \frac{10}{40} = \frac{4}{16} = \frac{8}{32} = \frac{3}{12} = \frac{12}{48} = \frac{2}{8} = \frac{9}{36} = \frac{5}{20} = \frac{7}{28} = \frac{14}{56}$

$\frac{1}{5} = \frac{7}{35} = \frac{3}{15} = \frac{9}{45} = \frac{4}{20} = \frac{6}{30} = \frac{13}{65} = \frac{5}{25} = \frac{14}{70} = \frac{10}{50} = \frac{2}{10} = \frac{8}{40}$

2. <u>Which</u> fraction is bigger ($\frac{white\ blocks}{blocks\ in\ total}$)?

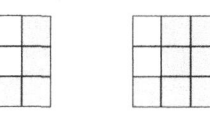

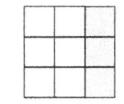

$\frac{2}{9} < \frac{4}{9} < \frac{6}{9}$

Page 18

1. <u>Find</u> an equivalent fraction with the given numerator.

Step 1: Use the given fraction ($\frac{1}{6}$).
Step 2: The numerator 1 has been multiplied by 2 to get 2.
Step 3: To get an equivalent fraction, the denominator 6 must also be multiplied by 2 to get 12.
Therefore $\frac{1}{6} = \frac{2}{12}$.

$\frac{1}{6} = \frac{2}{12} = \frac{9}{54} = \frac{4}{24} = \frac{12}{72} = \frac{6}{36} = \frac{10}{60} = \frac{8}{48} = \frac{3}{18} = \frac{7}{42} = \frac{11}{66} = \frac{5}{30}$

$\frac{1}{7} = \frac{2}{14} = \frac{7}{49} = \frac{12}{84} = \frac{5}{35} = \frac{9}{63} = \frac{3}{21} = \frac{8}{56} = \frac{6}{42} = \frac{10}{70} = \frac{11}{77} = \frac{4}{28}$

$\frac{1}{8} = \frac{2}{16} = \frac{6}{48} = \frac{4}{32} = \frac{9}{72} = \frac{3}{24} = \frac{11}{88} = \frac{8}{64} = \frac{5}{40} = \frac{10}{80} = \frac{7}{56} = \frac{12}{96}$

$\frac{1}{9} = \frac{5}{45} = \frac{3}{27} = \frac{9}{81} = \frac{2}{18} = \frac{11}{99} = \frac{7}{63} = \frac{12}{108} = \frac{4}{36} = \frac{10}{90} = \frac{6}{54} = \frac{8}{72}$

2. <u>Draw</u> and <u>color</u> the fraction of each shape.

(2 × 2) (2 × 4) (4 × 3)

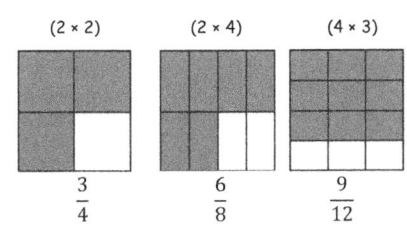

$\frac{3}{4}$ $\frac{6}{8}$ $\frac{9}{12}$

Page 19

1. <u>Find</u> an equivalent fraction with the given denominator.

Step 1: Use the given fraction ($\frac{1}{6}$).
Step 2: The denominator 6 has been multiplied by 3 to get 18.
Step 3: To get an equivalent fraction, the numerator 1 must also be multiplied by 3 to get 3.
Therefore $\frac{1}{6} = \frac{3}{18}$.

$\frac{1}{6} = \frac{3}{18} = \frac{6}{36} = \frac{2}{12} = \frac{8}{48} = \frac{11}{66} = \frac{4}{24} = \frac{13}{78} = \frac{7}{42} = \frac{10}{60} = \frac{5}{30} = \frac{9}{54}$

$\frac{1}{7} = \frac{3}{21} = \frac{7}{49} = \frac{10}{70} = \frac{2}{14} = \frac{11}{77} = \frac{6}{42} = \frac{4}{28} = \frac{9}{63} = \frac{5}{35} = \frac{12}{84} = \frac{8}{56}$

$\frac{1}{8} = \frac{3}{24} = \frac{5}{40} = \frac{2}{16} = \frac{11}{88} = \frac{9}{72} = \frac{6}{48} = \frac{8}{64} = \frac{10}{80} = \frac{12}{96} = \frac{4}{32} = \frac{7}{56}$

$\frac{1}{9} = \frac{4}{36} = \frac{2}{18} = \frac{8}{72} = \frac{3}{27} = \frac{10}{90} = \frac{5}{45} = \frac{7}{63} = \frac{11}{99} = \frac{6}{54} = \frac{9}{81} = \frac{13}{117}$

2. <u>Draw</u> and <u>color</u> the fraction of each shape.

(3 × 1) (3 × 3) (3 × 6)

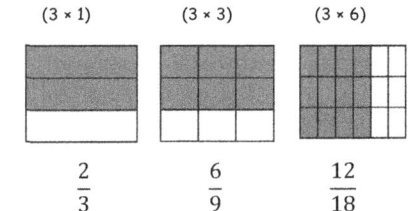

$\frac{2}{3}$ $\frac{6}{9}$ $\frac{12}{18}$

Page 20

About simplifying... When you divide both the numerator and denominator by the same number, you simplify the fraction, or write it in lower terms, or reduce it to its lowest terms.

I colored white the same relative amount of my block ($\frac{1}{4}$). Though the sizes of the equal parts my blocks are divided into are different.

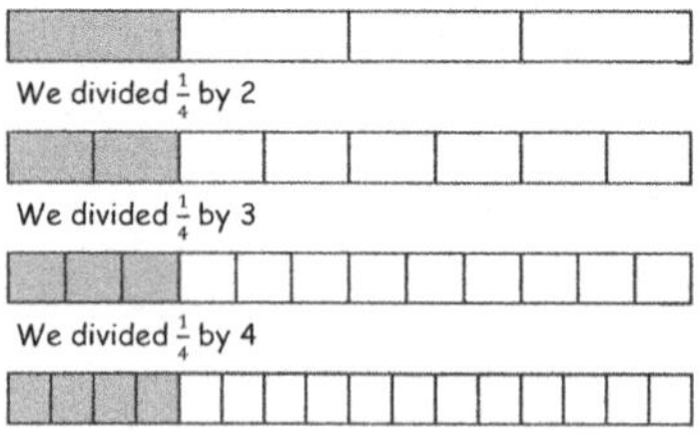

$\frac{1}{4} = \frac{1 \times 1}{4 \times 1}$

We divided $\frac{1}{4}$ by 2
$\frac{2}{8} = \frac{1 \times 2}{4 \times 2}$

We divided $\frac{1}{4}$ by 3
$\frac{3}{12} = \frac{1 \times 3}{4 \times 3}$

We divided $\frac{1}{4}$ by 4
$\frac{4}{16} = \frac{1 \times 4}{4 \times 4}$

All these fractions represent the same relative amount – one fourth.

The fraction $\frac{4}{16}$ is in higher terms, while $\frac{1}{4}$ is in its lowest terms. So $\frac{4}{16}$ was simplified or reduced to its lowest terms.

To simplify a fraction, divide both the numerator and the denominator by the same number and write a fraction in its simplest form:

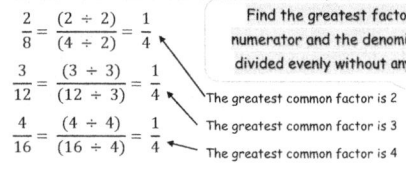

$\frac{2}{8} = \frac{(2 \div 2)}{(4 \div 2)} = \frac{1}{4}$ — The greatest common factor is 2

$\frac{3}{12} = \frac{(3 \div 3)}{(12 \div 3)} = \frac{1}{4}$ — The greatest common factor is 3

$\frac{4}{16} = \frac{(4 \div 4)}{(16 \div 4)} = \frac{1}{4}$ — The greatest common factor is 4

Find the greatest factor both the numerator and the denominator can be divided evenly without any remainder.

Page 21

1. Reduce each fraction to simplest form.

Well, to reduce a fraction to its lowest form (or term) means that I need to divide the numerator (the top number) and the denominator (the bottom number) by the SAME GREATEST number.

$\frac{(3 \div 3)}{(12 \div 3)} = \frac{1}{4}$

$\frac{3}{12} = \frac{(1 \times 3)}{(4 \times 3)} = \frac{1}{4}$ I prefer to say that I cancel the 3s.

So, you reduce the fraction to make it look as simple as possible. What is the GREATEST number both 2 and 4 can be divided into? Right, they both can be divided by 2 (the *greatest common factor*): $\frac{2}{4} = \frac{(2 \div 2)}{(4 \div 2)} = \frac{1}{2}$.

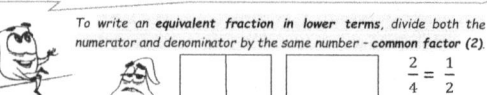

To write an *equivalent* fraction in lower terms, divide both the numerator and denominator by the same number – *common factor* (2).

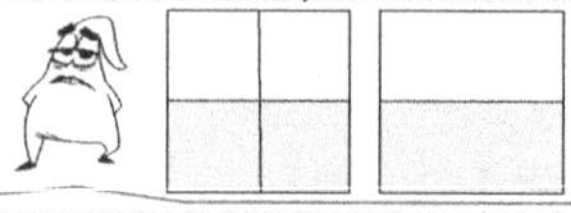

$\frac{2}{4} = \frac{1}{2}$

$\frac{(2 \div 2)}{(4 \div 2)} = \frac{1}{2}$

I can divide both the numerator and the denominator by 2, 3, 4, 5, 6, and etc. I need to reduce $\frac{18}{42}$ to its lowest terms: $\frac{(18 \div 2)}{(42 \div 2)} = \frac{(9 \div 3)}{(21 \div 3)} = \frac{3}{7}$.

I don't change the VALUE of the fraction. $\frac{(16 \div 2)}{(40 \div 2)} = \frac{(8 \div 2)}{(20 \div 2)} = \frac{(4 \div 2)}{(10 \div 2)} = \frac{2}{5}$

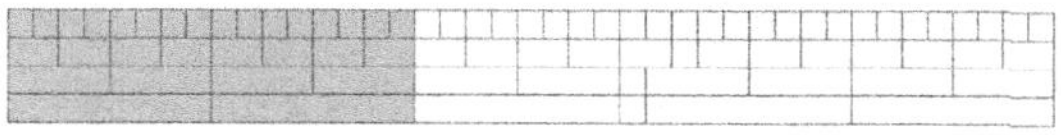

Page 22

1. <u>Write</u> each fraction in its simplest form. <u>Write</u> the missing numbers.

$\frac{(3 \div 3)}{(6 \div 3)} = \frac{1}{2}$ $\frac{(3 \div 3)}{(12 \div 3)} = \frac{1}{4}$

The Greatest Common factor = 3   *The Greatest Common factor = 3*

$\frac{(2 \div 2)}{(14 \div 2)} = \frac{1}{7}$ $\frac{(2 \div 2)}{(4 \div 2)} = \frac{1}{2}$

The Greatest Common factor = 2   *The Greatest Common factor = 2*

$\frac{(5 \div 5)}{(20 \div 5)} = \frac{1}{4}$ $\frac{(5 \div 5)}{(35 \div 5)} = \frac{1}{7}$

The Greatest Common factor = 5   *The Greatest Common factor = 5*

$\frac{(4 \div 4)}{(24 \div 4)} = \frac{1}{6}$ $\frac{(4 \div 4)}{(32 \div 4)} = \frac{1}{8}$

The Greatest Common factor = 4   *The Greatest Common factor = 4*

$\frac{(6 \div 6)}{(18 \div 6)} = \frac{1}{3}$ $\frac{(6 \div 6)}{(42 \div 6)} = \frac{1}{7}$

The Greatest Common factor = 6   *The Greatest Common factor = 6*

$\frac{(7 \div 7)}{(14 \div 7)} = \frac{1}{2}$ $\frac{(7 \div 7)}{(28 \div 7)} = \frac{1}{4}$

The Greatest Common factor = 7   *The Greatest Common factor = 7*

$\frac{(8 \div 8)}{(48 \div 8)} = \frac{1}{6}$ $\frac{(8 \div 8)}{(32 \div 8)} = \frac{1}{4}$

The Greatest Common factor = 8   *The Greatest Common factor = 8*

2. <u>Compare</u> fractions using ">", "<", or "=".

$\frac{3}{8} < \frac{5}{8}$ $\frac{4}{12} < \frac{9}{12}$

$\frac{5}{7} > \frac{2}{7}$ $\frac{3}{9} < \frac{7}{9}$

Page 23

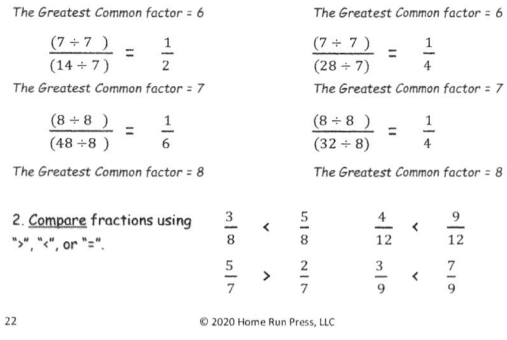

1. <u>Reduce</u> each fraction to its lowest terms and <u>color</u> the fraction of each shape. <u>Write</u> the missing numbers.

          Blue   Green   Red

$\frac{16}{28} = \frac{(16 \div 2)}{(28 \div 2)} = \frac{(8 \div 2)}{(14 \div 2)} = \frac{4}{7}$ *The Greatest common factor is the greatest factor that divides both the numerator and the denominator evenly into a number, and there is no remainder.*

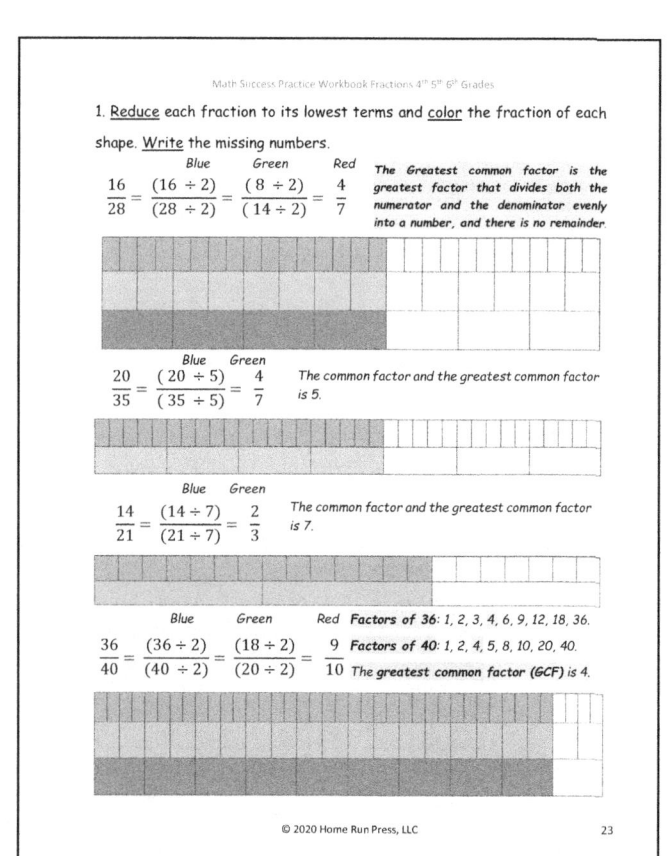

        Blue   Green

$\frac{20}{35} = \frac{(20 \div 5)}{(35 \div 5)} = \frac{4}{7}$ *The common factor and the greatest common factor is 5.*

        Blue   Green

$\frac{14}{21} = \frac{(14 \div 7)}{(21 \div 7)} = \frac{2}{3}$ *The common factor and the greatest common factor is 7.*

        Blue   Green   Red   **Factors of 36**: 1, 2, 3, 4, 6, 9, 12, 18, 36.

$\frac{36}{40} = \frac{(36 \div 2)}{(40 \div 2)} = \frac{(18 \div 2)}{(20 \div 2)} = \frac{9}{10}$ **Factors of 40**: 1, 2, 4, 5, 8, 10, 20, 40.
*The **greatest common factor** (GCF) is 4.*

Page 24

1. <u>Reduce</u> each fraction to its simplest form and <u>color</u> the fraction of each shape. <u>Write</u> the missing numbers.

Blue   Green   Red   Yellow $2 \times 3 \times 3 = 18$

$\dfrac{18}{36} = \dfrac{(18 \div 2)}{(36 \div 2)} = \dfrac{(9 \div 3)}{(18 \div 3)} = \dfrac{(3 \div 3)}{(6 \div 3)} = \dfrac{1}{2}$ *18 is the greatest common factor*

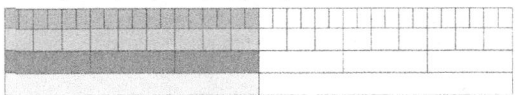

Blue   Green   Red   Yellow $2 \times 2 \times 2 = 8$

$\dfrac{32}{40} = \dfrac{(32 \div 2)}{(40 \div 2)} = \dfrac{(16 \div 2)}{(20 \div 2)} = \dfrac{(8 \div 2)}{(10 \div 2)} = \dfrac{4}{5}$ *8 is the greatest common factor*

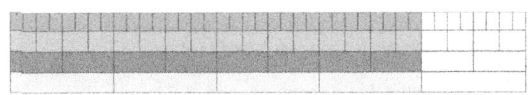

Blue   Green   Red $2 \times 2 = 4$

$\dfrac{28}{36} = \dfrac{(28 \div 2)}{(36 \div 2)} = \dfrac{(14 \div 2)}{(18 \div 2)} = \dfrac{7}{9}$ *4 is the greatest common factor*

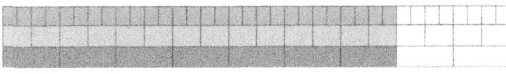

Blue   Yellow

$\dfrac{15}{18} = \dfrac{(15 \div 3)}{(18 \div 3)} = \dfrac{5}{6}$ *3 is the greatest common factor*

Page 25

1. <u>Reduce</u> each fraction to its simplest form. <u>Write</u> the missing numbers.

$\dfrac{32}{48} = \dfrac{(32 \div 2)}{(48 \div 2)} = \dfrac{(16 \div 2)}{(24 \div 2)} = \dfrac{(8 \div 2)}{(12 \div 2)} = \dfrac{(4 \div 2)}{(6 \div 2)} = \dfrac{2}{3}$

$\dfrac{(32 \div 16)}{(48 \div 16)} = \dfrac{2}{3}$ $2 \times 2 \times 2 \times 2 = 16$ *is the greatest common factor*

$\dfrac{12}{28} = \dfrac{(12 \div 2)}{(28 \div 2)} = \dfrac{(6 \div 2)}{(14 \div 2)} = \dfrac{3}{7}$ $\dfrac{(12 \div 4)}{(28 \div 4)} = \dfrac{3}{7}$

$2 \times 2 = 4$ *is the greatest common factor*

$\dfrac{36}{48} = \dfrac{(36 \div 2)}{(48 \div 2)} = \dfrac{(18 \div 2)}{(24 \div 2)} = \dfrac{(9 \div 3)}{(12 \div 3)} = \dfrac{3}{4}$ $\dfrac{(36 \div 12)}{(48 \div 12)} = \dfrac{3}{4}$

$2 \times 2 \times 3 = 12$ *is the greatest common factor*

$\dfrac{50}{75} = \dfrac{(50 \div 5)}{(75 \div 5)} = \dfrac{(10 \div 5)}{(15 \div 5)} = \dfrac{2}{3}$ $\dfrac{(50 \div 25)}{(75 \div 25)} = \dfrac{2}{3}$

$5 \times 5 = 25$ *is the greatest common factor*

2. <u>Write</u> the fraction name under each picture. <u>Compare</u> the fractions.

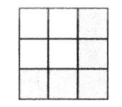

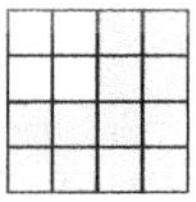

$\dfrac{white\ blocks}{blocks\ in\ total}$ $\dfrac{4}{9}$ > $\dfrac{4}{16}$

Page 26

1. <u>Reduce</u> each fraction to its simplest form. <u>Write</u> the missing numbers.

$\dfrac{56}{72} = \dfrac{(56 \div 2)}{(72 \div 2)} = \dfrac{(28 \div 2)}{(36 \div 2)} = \dfrac{(14 \div 2)}{(18 \div 2)} = \dfrac{7}{9}$ $\dfrac{(56 \div 8)}{(72 \div 8)} = \dfrac{7}{9}$

$2 \times 2 \times 2 = 8$ *is the greatest common factor*

$\dfrac{36}{81} = \dfrac{(36 \div 3)}{(81 \div 3)} = \dfrac{(12 \div 3)}{(27 \div 3)} = \dfrac{4}{9}$ $\dfrac{(36 \div 9)}{(81 \div 9)} = \dfrac{4}{9}$

$3 \times 3 = 9$ *is the greatest common factor*

$\dfrac{60}{96} = \dfrac{(60 \div 2)}{(96 \div 2)} = \dfrac{(30 \div 2)}{(48 \div 2)} = \dfrac{(15 \div 3)}{(24 \div 2)} = \dfrac{5}{8}$ $\dfrac{(60 \div 12)}{(96 \div 12)} = \dfrac{5}{8}$

$2 \times 2 \times 3 = 12$ *is the greatest common factor*

$\dfrac{48}{96} = \dfrac{(48 \div 2)}{(96 \div 2)} = \dfrac{(24 \div 2)}{(48 \div 2)} = \dfrac{(12 \div 2)}{(24 \div 2)} = \dfrac{(6 \div 2)}{(12 \div 2)} = \dfrac{(3 \div 3)}{(6 \div 3)} = \dfrac{1}{2}$

$\dfrac{(48 \div 48)}{(96 \div 48)} = \dfrac{1}{2}$ $2 \times 2 \times 2 \times 3 = 48$ *is the greatest common factor*

2. <u>Write</u> the fraction name under each picture. <u>Compare</u> fractions.

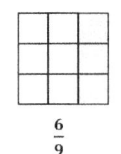

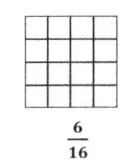

$\dfrac{6}{9}$ > $\dfrac{6}{16}$

Page 27

1. <u>Reduce</u> each fraction to its simplest form. <u>Find</u> the greatest common factor (GCF).

$\dfrac{(20 \div 4)}{(28 \div 4)} = \dfrac{5}{7}$ $\dfrac{(18 \div 9)}{(27 \div 9)} = \dfrac{2}{3}$ $\dfrac{(12 \div 6)}{(42 \div 6)} = \dfrac{2}{7}$

GCF = 4   GCF = 9   GCF = 6

$\dfrac{(6 \div 3)}{(63 \div 3)} = \dfrac{2}{21}$ $\dfrac{(15 \div 5)}{(35 \div 5)} = \dfrac{3}{7}$ $\dfrac{(49 \div 7)}{(77 \div 7)} = \dfrac{7}{11}$

GCF = 3   GCF = 5   GCF = 7

$\dfrac{(24 \div 8)}{(56 \div 8)} = \dfrac{3}{7}$ $\dfrac{(16 \div 4)}{(36 \div 4)} = \dfrac{4}{9}$ $\dfrac{(20 \div 4)}{(32 \div 4)} = \dfrac{5}{8}$

GCF = 8   GCF = 4   GCF = 4

$\dfrac{(10 \div 2)}{(34 \div 2)} = \dfrac{5}{17}$ $\dfrac{(21 \div 3)}{(60 \div 3)} = \dfrac{7}{20}$ $\dfrac{(9 \div 3)}{(24 \div 3)} = \dfrac{3}{8}$

GCF = 2   GCF = 3   GCF = 3

$\dfrac{(25 \div 5)}{(45 \div 5)} = \dfrac{5}{9}$ $\dfrac{(40 \div 8)}{(64 \div 8)} = \dfrac{5}{8}$ $\dfrac{(35 \div 7)}{(84 \div 7)} = \dfrac{5}{12}$

GCF = 5   GCF = 8   GCF = 7

2. <u>Color</u> the fraction of each shape. <u>Compare</u> fractions.

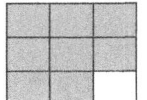

$\dfrac{8}{9}$ > $\dfrac{8}{12}$

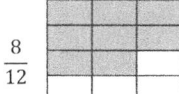

127

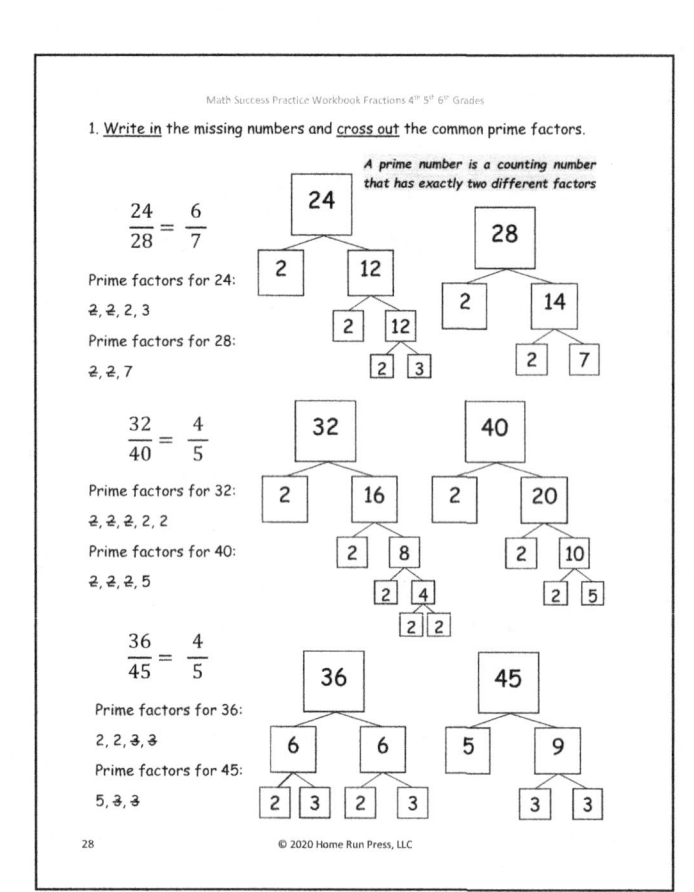

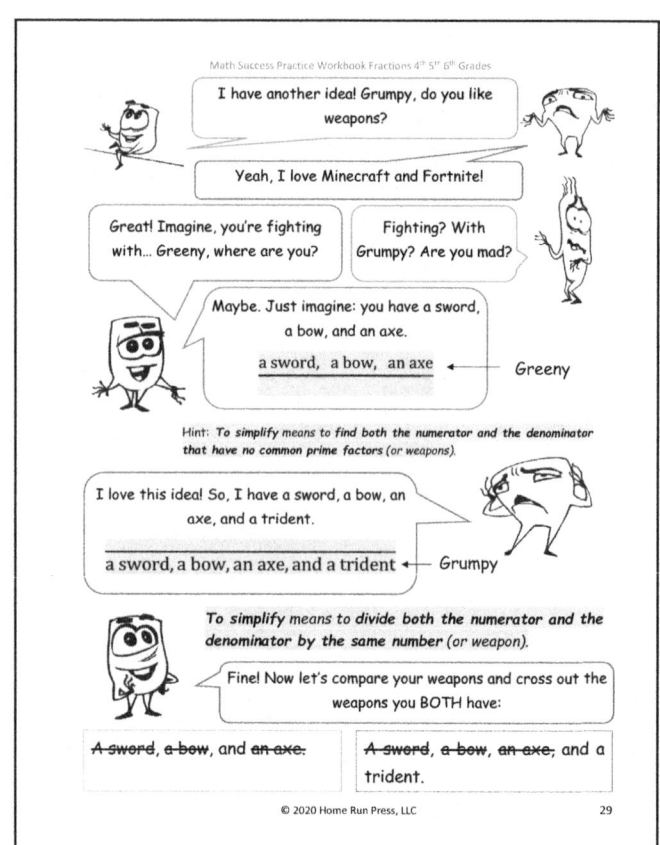

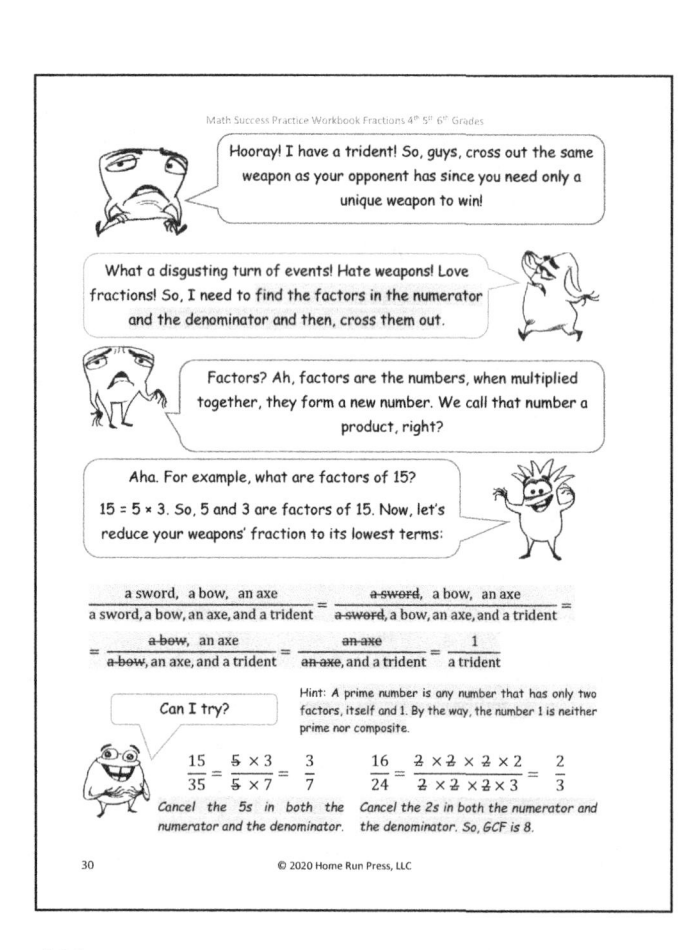

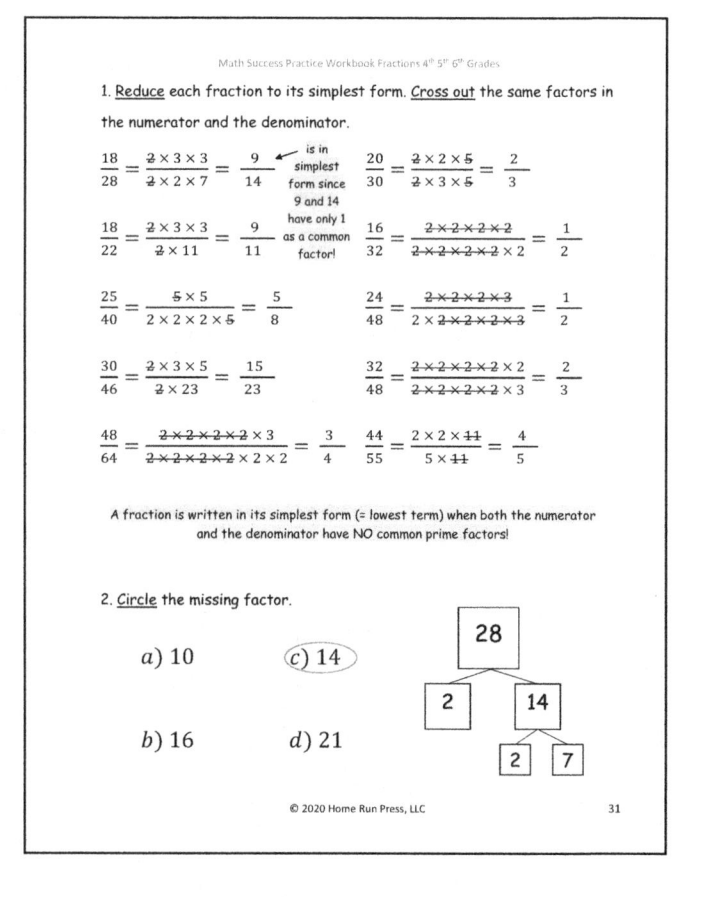

1. <u>Write in</u> the missing numbers and <u>cross out</u> the common prime factors.

A prime number is a counting number that has exactly two different factors

$\frac{24}{28} = \frac{6}{7}$

Prime factors for 24:
~~2, 2,~~ 2, 3

Prime factors for 28:
~~2, 2,~~ 7

24 → 2, 12 → 2, 12 → 2, 3
28 → 2, 14 → 2, 7

$\frac{32}{40} = \frac{4}{5}$

Prime factors for 32:
~~2, 2, 2,~~ 2, 2

Prime factors for 40:
~~2, 2, 2,~~ 5

32 → 2, 16 → 2, 8 → 2, 4 → 2, 2
40 → 2, 20 → 2, 10 → 2, 5

$\frac{36}{45} = \frac{4}{5}$

Prime factors for 36:
2, 2, ~~3, 3~~

Prime factors for 45:
5, ~~3, 3~~

36 → 6, 6 → 2, 3, 2, 3
45 → 5, 9 → 3, 3

I have another idea! Grumpy, do you like weapons?

Yeah, I love Minecraft and Fortnite!

Great! Imagine, you're fighting with... Greeny, where are you?

Fighting? With Grumpy? Are you mad?

Maybe. Just imagine: you have a sword, a bow, and an axe.

a sword, a bow, an axe ← Greeny

Hint: *To simplify means to find both the numerator and the denominator that have no common prime factors (or weapons).*

I love this idea! So, I have a sword, a bow, an axe, and a trident.

a sword, a bow, an axe, and a trident ← Grumpy

To simplify means to divide both the numerator and the denominator by the same number (or weapon).

Fine! Now let's compare your weapons and cross out the weapons you BOTH have:

~~A sword~~, ~~a bow~~, and ~~an axe~~.

~~A sword~~, ~~a bow~~, ~~an axe~~, and a trident.

Hooray! I have a trident! So, guys, cross out the same weapon as your opponent has since you need only a unique weapon to win!

What a disgusting turn of events! Hate weapons! Love fractions! So, I need to find the factors in the numerator and the denominator and then, cross them out.

Factors? Ah, factors are the numbers, when multiplied together, they form a new number. We call that number a product, right?

Aha. For example, what are factors of 15?

15 = 5 × 3. So, 5 and 3 are factors of 15. Now, let's reduce your weapons' fraction to its lowest terms:

$\frac{\text{a sword, a bow, an axe}}{\text{a sword, a bow, an axe, and a trident}} = \frac{\text{~~a sword~~, a bow, an axe}}{\text{~~a sword~~, a bow, an axe, and a trident}} =$

$= \frac{\text{~~a bow~~, an axe}}{\text{~~a bow~~, an axe, and a trident}} = \frac{\text{~~an axe~~}}{\text{~~an axe~~, and a trident}} = \frac{1}{\text{a trident}}$

Can I try?

Hint: *A prime number is any number that has only two factors, itself and 1. By the way, the number 1 is neither prime nor composite.*

$\frac{15}{35} = \frac{\cancel{5} \times 3}{\cancel{5} \times 7} = \frac{3}{7}$

Cancel the 5s in both the numerator and the denominator.

$\frac{16}{24} = \frac{\cancel{2} \times \cancel{2} \times \cancel{2} \times 2}{\cancel{2} \times \cancel{2} \times \cancel{2} \times 3} = \frac{2}{3}$

Cancel the 2s in both the numerator and the denominator. So, GCF is 8.

---

1. <u>Reduce</u> each fraction to its simplest form. <u>Cross out</u> the same factors in the numerator and the denominator.

$\frac{18}{28} = \frac{\cancel{2} \times 3 \times 3}{\cancel{2} \times 2 \times 7} = \frac{9}{14}$ ← is in simplest form since 9 and 14 have only 1 as a common factor!

$\frac{20}{30} = \frac{\cancel{2} \times 2 \times \cancel{5}}{\cancel{2} \times 3 \times \cancel{5}} = \frac{2}{3}$

$\frac{18}{22} = \frac{\cancel{2} \times 3 \times 3}{\cancel{2} \times 11} = \frac{9}{11}$

$\frac{16}{32} = \frac{\cancel{2} \times \cancel{2} \times \cancel{2} \times 2}{\cancel{2} \times \cancel{2} \times \cancel{2} \times 2 \times 2} = \frac{1}{2}$

$\frac{25}{40} = \frac{\cancel{5} \times 5}{2 \times 2 \times 2 \times \cancel{5}} = \frac{5}{8}$

$\frac{24}{48} = \frac{\cancel{2} \times \cancel{2} \times \cancel{2} \times 3}{2 \times \cancel{2} \times \cancel{2} \times \cancel{2} \times 3} = \frac{1}{2}$

$\frac{30}{46} = \frac{\cancel{2} \times 3 \times 5}{\cancel{2} \times 23} = \frac{15}{23}$

$\frac{32}{48} = \frac{\cancel{2} \times \cancel{2} \times \cancel{2} \times \cancel{2} \times 2}{\cancel{2} \times \cancel{2} \times \cancel{2} \times \cancel{2} \times 3} = \frac{2}{3}$

$\frac{48}{64} = \frac{\cancel{2} \times \cancel{2} \times \cancel{2} \times \cancel{2} \times 3}{\cancel{2} \times \cancel{2} \times \cancel{2} \times \cancel{2} \times 2 \times 2} = \frac{3}{4}$

$\frac{44}{55} = \frac{2 \times 2 \times \cancel{11}}{5 \times \cancel{11}} = \frac{4}{5}$

A fraction is written in its simplest form (= lowest term) when both the numerator and the denominator have NO common prime factors!

2. <u>Circle</u> the missing factor.

a) 10   (c) 14)

b) 16   d) 21

28 → 2, 14 → 2, 7

Page 32

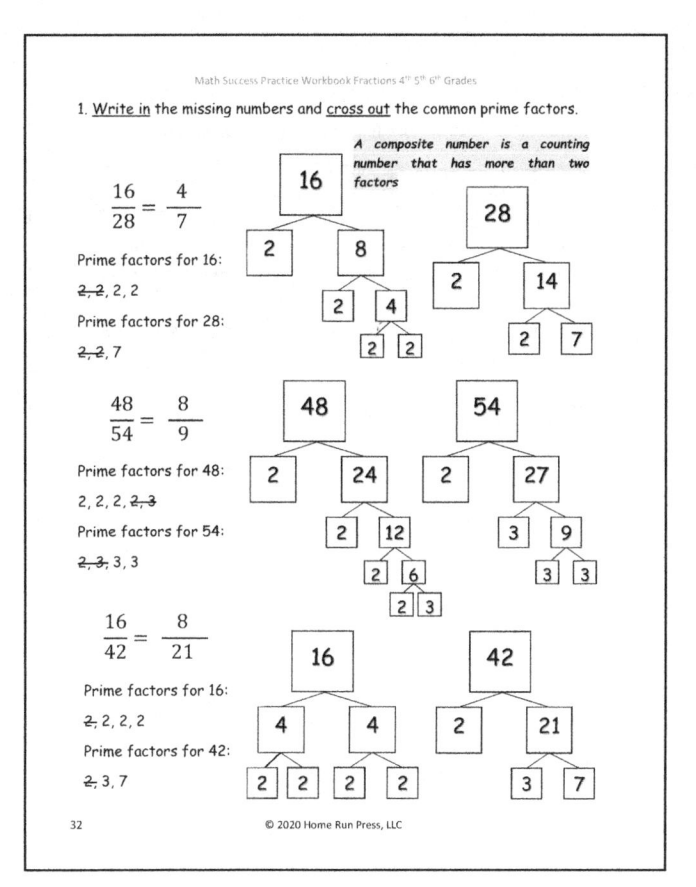

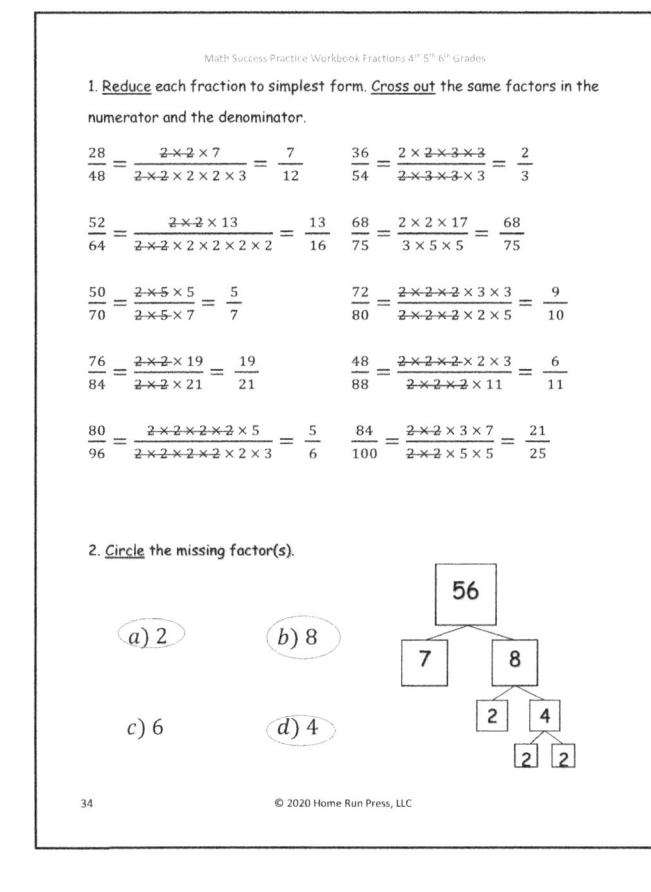

Math Success Practice Workbook Fractions 4th 5th 6th Grades

1. <u>Write in</u> the missing numbers and <u>cross out</u> the common prime factors.

A composite number is a counting number that has more than two factors

$\dfrac{16}{28} = \dfrac{4}{7}$

Prime factors for 16:
~~2,~~ ~~2,~~ 2, 2

Prime factors for 28:
~~2,~~ ~~2,~~ 7

Tree for 16: 16 → 2, 8; 8 → 2, 4; 4 → 2, 2
Tree for 28: 28 → 2, 14; 14 → 2, 7

$\dfrac{48}{54} = \dfrac{8}{9}$

Prime factors for 48:
2, 2, 2, ~~2, 3~~

Prime factors for 54:
~~2, 3,~~ 3, 3

Tree for 48: 48 → 2, 24; 24 → 2, 12; 12 → 2, 6; 6 → 2, 3
Tree for 54: 54 → 2, 27; 27 → 3, 9; 9 → 3, 3

$\dfrac{16}{42} = \dfrac{8}{21}$

Prime factors for 16:
~~2,~~ 2, 2, 2

Prime factors for 42:
~~2,~~ 3, 7

Tree for 16: 16 → 4, 4; 4 → 2, 2; 4 → 2, 2
Tree for 42: 42 → 2, 21; 21 → 3, 7

Page 33

1. <u>Write in</u> the missing numbers.

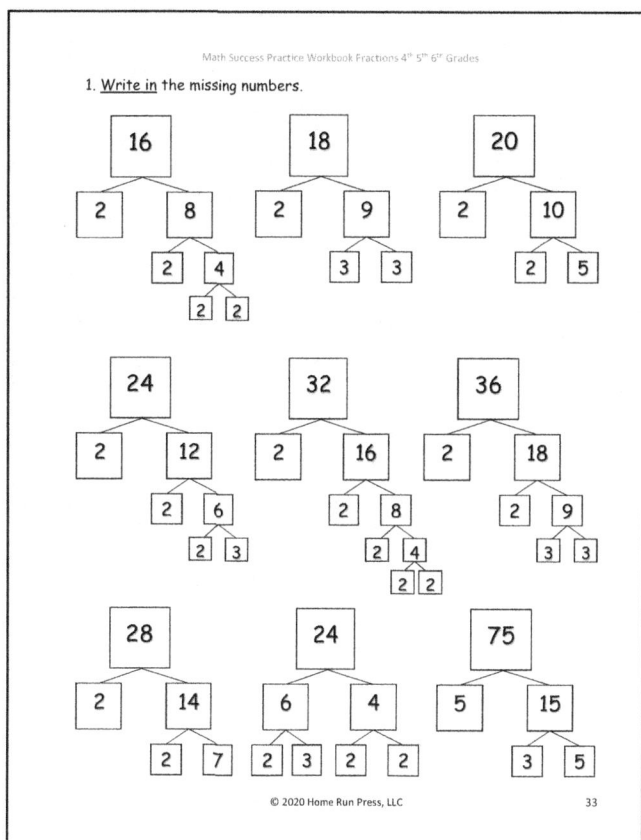

Tree: 16 → 2, 8; 8 → 2, 4; 4 → 2, 2
Tree: 18 → 2, 9; 9 → 3, 3
Tree: 20 → 2, 10; 10 → 2, 5

Tree: 24 → 2, 12; 12 → 2, 6; 6 → 2, 3
Tree: 32 → 2, 16; 16 → 2, 8; 8 → 2, 4; 4 → 2, 2
Tree: 36 → 2, 18; 18 → 2, 9; 9 → 3, 3

Tree: 28 → 2, 14; 14 → 2, 7
Tree: 24 → 6, 4; 6 → 2, 3; 4 → 2, 2
Tree: 75 → 5, 15; 15 → 3, 5

Page 34

1. <u>Reduce</u> each fraction to simplest form. <u>Cross out</u> the same factors in the numerator and the denominator.

$\dfrac{28}{48} = \dfrac{\cancel{2\times 2}\times 7}{\cancel{2\times 2}\times 2\times 2\times 3} = \dfrac{7}{12}$

$\dfrac{36}{54} = \dfrac{2\times \cancel{2}\times \cancel{3}\times \cancel{3}}{\cancel{2}\times \cancel{3}\times \cancel{3}\times 3} = \dfrac{2}{3}$

$\dfrac{52}{64} = \dfrac{\cancel{2\times 2}\times 13}{\cancel{2\times 2}\times 2\times 2\times 2\times 2} = \dfrac{13}{16}$

$\dfrac{68}{75} = \dfrac{2\times 2\times 17}{3\times 5\times 5} = \dfrac{68}{75}$

$\dfrac{50}{70} = \dfrac{\cancel{2\times 5}\times 5}{\cancel{2\times 5}\times 7} = \dfrac{5}{7}$

$\dfrac{72}{80} = \dfrac{\cancel{2\times 2\times 2}\times 3\times 3}{\cancel{2\times 2\times 2}\times 2\times 5} = \dfrac{9}{10}$

$\dfrac{76}{84} = \dfrac{\cancel{2\times 2}\times 19}{\cancel{2\times 2}\times 21} = \dfrac{19}{21}$

$\dfrac{48}{88} = \dfrac{\cancel{2\times 2}\times 2\times 2\times 3}{\cancel{2\times 2}\times 2\times 11} = \dfrac{6}{11}$

$\dfrac{80}{96} = \dfrac{\cancel{2\times 2\times 2\times 2}\times 5}{\cancel{2\times 2\times 2\times 2}\times 2\times 3} = \dfrac{5}{6}$

$\dfrac{84}{100} = \dfrac{\cancel{2\times 2}\times 3\times 7}{\cancel{2\times 2}\times 5\times 5} = \dfrac{21}{25}$

2. <u>Circle</u> the missing factor(s).

Tree: 56 → 7, 8; 8 → 2, 4; 4 → 2, 2

a) 2 (circled)   b) 8 (circled)
c) 6   d) 4 (circled)

Page 35

1. <u>Write in</u> the missing numbers.

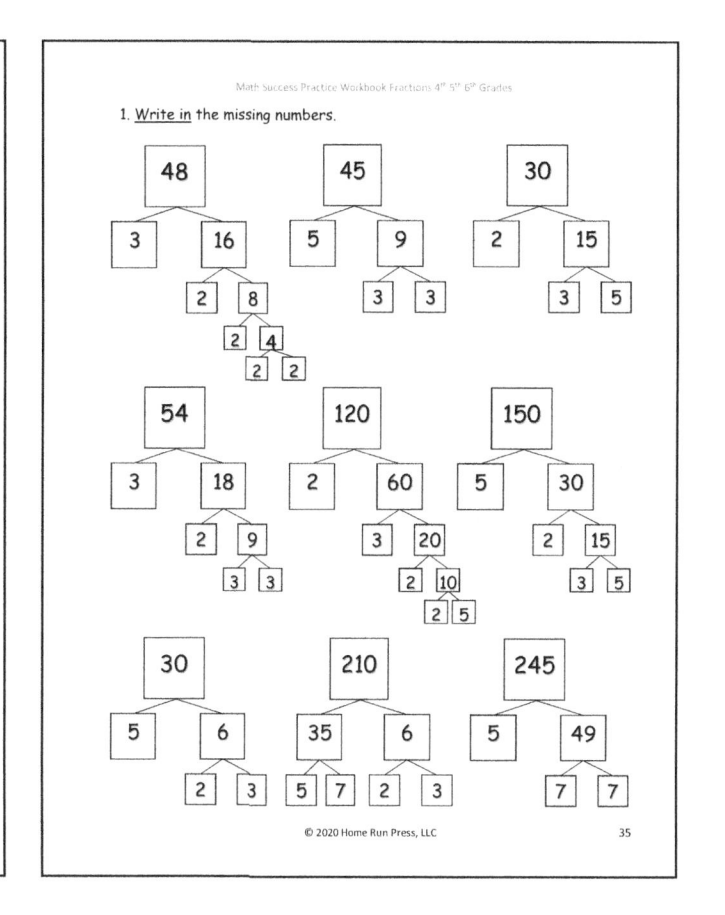

Tree: 48 → 3, 16; 16 → 2, 8; 8 → 2, 4; 4 → 2, 2
Tree: 45 → 5, 9; 9 → 3, 3
Tree: 30 → 2, 15; 15 → 3, 5

Tree: 54 → 3, 18; 18 → 2, 9; 9 → 3, 3
Tree: 120 → 2, 60; 60 → 3, 20; 20 → 2, 10; 10 → 2, 5
Tree: 150 → 5, 30; 30 → 2, 15; 15 → 3, 5

Tree: 30 → 5, 6; 6 → 2, 3
Tree: 210 → 6, 35; 6 → 2, 3; 35 → 5, 7
Tree: 245 → 5, 49; 49 → 7, 7

129

Page 36

1. <u>Compare</u> fractions using "<," ">," or "=."

Help! SOS! Help!!!

Quick, Smarty, help him! I found a chest with diamonds! I want to play!

Congratulations, Grumpy! Purple-y, look, to compare fractions means to find out which fraction and smaller or larger.

When fractions have the same denominator, the fraction with the greatest (largest) numerator has the greatest value.

6 out of 16 squares are less than 9 out of 16 squares.

$$\frac{6}{16} < \frac{9}{16}$$

Like denominators

I compare the unlike fractions (= fractions with the different numerators or denominators). Let's start with fractions with like numerators but with unlike denominators:

$$\frac{2}{4} \quad \frac{2}{16}$$

Unlike denominators

I divided my block into 4 huge equal parts and colored 2 parts white. Then, I divided another block into 16 tiny equal parts and colored 2 equal parts white. They look different!

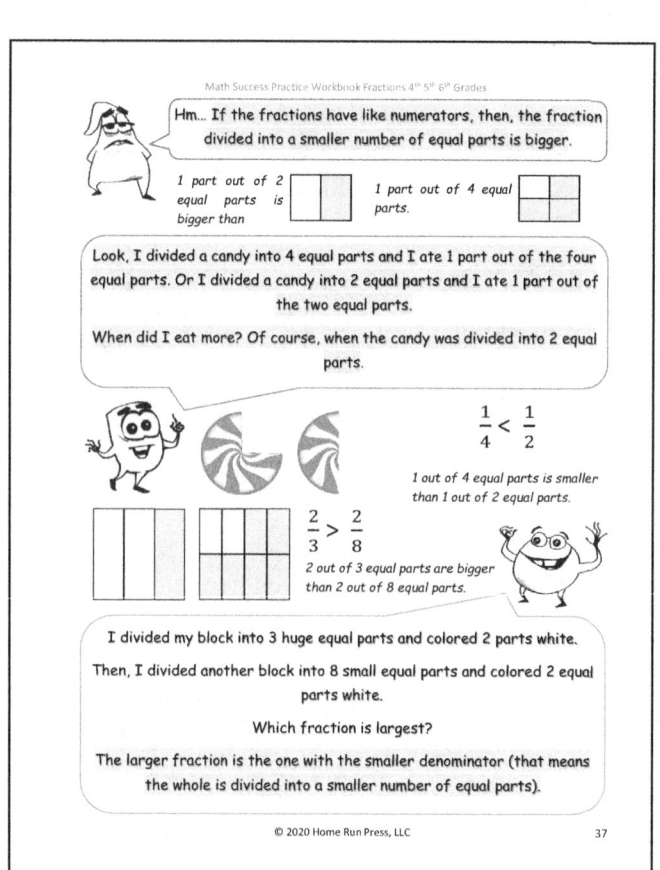

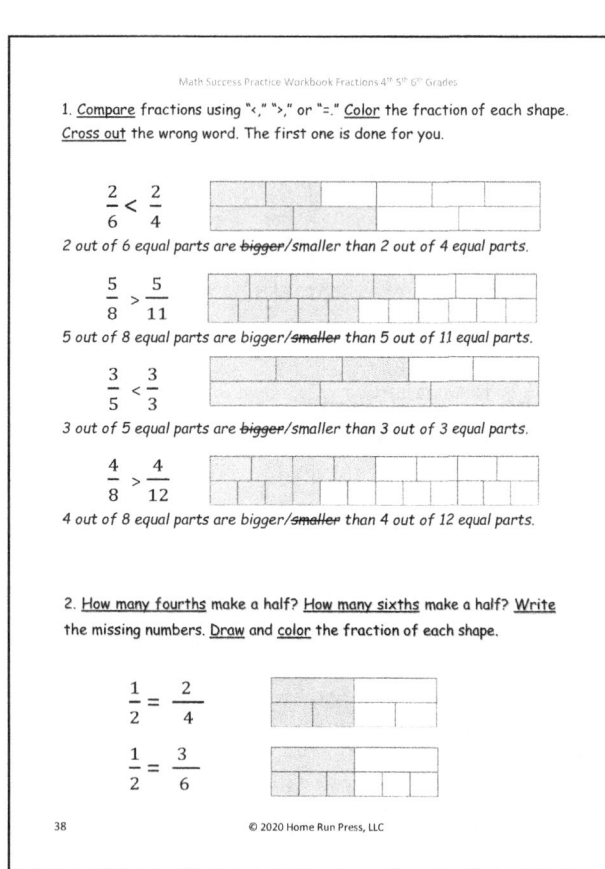

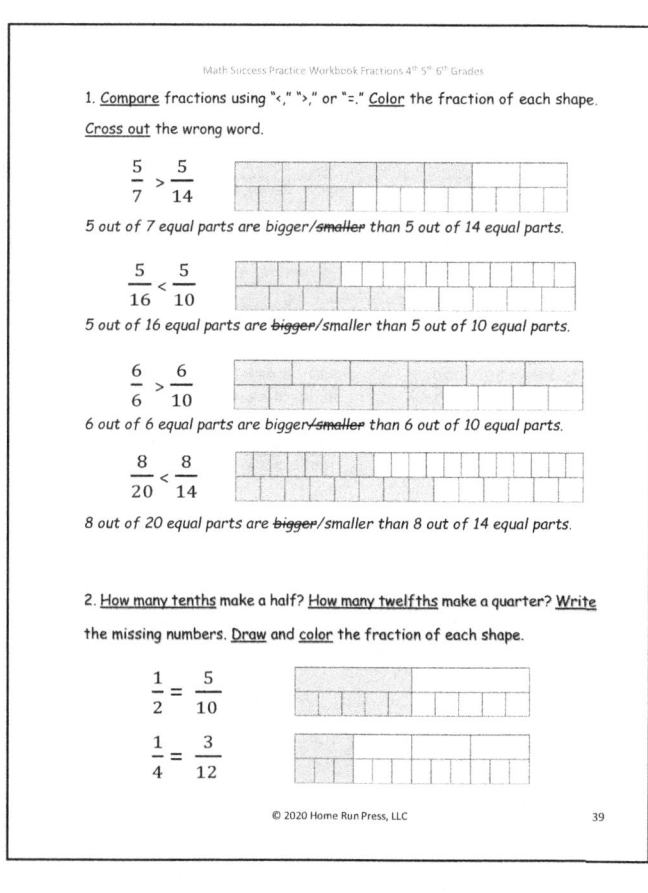

Page 37

Hm... If the fractions have like numerators, then, the fraction divided into a smaller number of equal parts is bigger.

1 part out of 2 equal parts is bigger than

1 part out of 4 equal parts.

Look, I divided a candy into 4 equal parts and I ate 1 part out of the four equal parts. Or I divided a candy into 2 equal parts and I ate 1 part out of the two equal parts.

When did I eat more? Of course, when the candy was divided into 2 equal parts.

$$\frac{1}{4} < \frac{1}{2}$$

1 out of 4 equal parts is smaller than 1 out of 2 equal parts.

$$\frac{2}{3} > \frac{2}{8}$$

2 out of 3 equal parts are bigger than 2 out of 8 equal parts.

I divided my block into 3 huge equal parts and colored 2 parts white.

Then, I divided another block into 8 small equal parts and colored 2 equal parts white.

Which fraction is largest?

The larger fraction is the one with the smaller denominator (that means the whole is divided into a smaller number of equal parts).

Page 38

1. <u>Compare</u> fractions using "<," ">," or "=." <u>Color</u> the fraction of each shape. <u>Cross out</u> the wrong word. The first one is done for you.

$$\frac{2}{6} < \frac{2}{4}$$

2 out of 6 equal parts are ~~bigger~~/smaller than 2 out of 4 equal parts.

$$\frac{5}{8} > \frac{5}{11}$$

5 out of 8 equal parts are bigger/~~smaller~~ than 5 out of 11 equal parts.

$$\frac{3}{5} < \frac{3}{3}$$

3 out of 5 equal parts are ~~bigger~~/smaller than 3 out of 3 equal parts.

$$\frac{4}{8} > \frac{4}{12}$$

4 out of 8 equal parts are bigger/~~smaller~~ than 4 out of 12 equal parts.

2. <u>How many fourths</u> make a half? <u>How many sixths</u> make a half? <u>Write</u> the missing numbers. <u>Draw</u> and <u>color</u> the fraction of each shape.

$$\frac{1}{2} = \frac{2}{4}$$

$$\frac{1}{2} = \frac{3}{6}$$

Page 39

1. <u>Compare</u> fractions using "<," ">," or "=." <u>Color</u> the fraction of each shape. <u>Cross out</u> the wrong word.

$$\frac{5}{7} > \frac{5}{14}$$

5 out of 7 equal parts are bigger/~~smaller~~ than 5 out of 14 equal parts.

$$\frac{5}{16} < \frac{5}{10}$$

5 out of 16 equal parts are ~~bigger~~/smaller than 5 out of 10 equal parts.

$$\frac{6}{6} > \frac{6}{10}$$

6 out of 6 equal parts are bigger/~~smaller~~ than 6 out of 10 equal parts.

$$\frac{8}{20} < \frac{8}{14}$$

8 out of 20 equal parts are ~~bigger~~/smaller than 8 out of 14 equal parts.

2. <u>How many tenths</u> make a half? <u>How many twelfths</u> make a quarter? <u>Write</u> the missing numbers. <u>Draw</u> and <u>color</u> the fraction of each shape.

$$\frac{1}{2} = \frac{5}{10}$$

$$\frac{1}{4} = \frac{3}{12}$$

Page 40

I'm a Minecraft fan. So, Steve meets $\frac{1}{3}$ of as many zombies as Alex. If Alex meets 12 zombies, how many zombies does Steve meet?

You need to find parts of a number. Remember that the word OF in fractions means MULTIPLY. So, Steve meets $\frac{1}{3}$ of 12 zombies in Math means $\frac{1}{3} \times 12$.

I have an idea! I need to divide my rectangle into 12 equal parts and color $\frac{1}{3}$ of them brown.

Step 1: I find an equivalent fraction with the denominator 12: $\frac{1}{3} = \frac{4}{12}$ (remember, equivalent fractions are fractions that represent the same value).

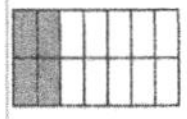

Step 2: The numerator shows how many parts I need to color brown. I need to color 4 parts brown since 4 parts out of 12 equal parts is the same as 1 part out of 3 equal parts.

$\frac{1}{3} = \frac{4}{12}$

Another algorithm:

Step 1: OF means multiply: $\frac{1}{3} \times 12$

Step 2: write 12 as $\frac{12}{1}$: $\frac{1}{3} \times \frac{12}{1}$

Step 3: Multiply numerators and denominators, cancel the 3s:

$\frac{1}{3} \times 12 = \frac{1}{3} \times \frac{12}{1} = \frac{1 \times 12}{3 \times 1} = \frac{1 \times \cancel{12}^4}{\cancel{3} \times 1} = \frac{1 \times 4}{1 \times 1} = 4$

Page 41

Aha... I have another idea! All the zombies are divided equally into 3 equal groups. And I meet only $\frac{1}{3}$ of them.

The question is: If there are 12 zombies altogether, how many zombies are in one of the 3 equal parts?

Step 1: find how many zombies are in one of the three equal parts: $12 \div 3 = 4$

(4 zombies are in each of the three equal groups)

Step 2: find how many zombies are in one group: $1 \times 4 = 4$

(4 zombies in 1 group).

I wonder, if I meet $\frac{2}{3}$ of 12 zombies, how many zombies do I meet?

That's the question I like. My favorite algorithm:

$\frac{2}{3} \times 12 = \frac{2}{3} \times \frac{12}{1} = \frac{2 \times 12}{3 \times 1} = \frac{2 \times \cancel{12}}{\cancel{3} \times 1} = \frac{2 \times 4}{1 \times 1} = 8$ (zombies)

I like to solve it slower: $\frac{2}{3} \times 12$

Step 1: $12 \div 3 = 4$ (find $\frac{1}{3}$ out of 12)

(4 zombies in each of the three equal groups)

Step 2: $2 \times 4 = 8$ (find how many zombies are in 2 groups)

(8 zombies in 2 groups).

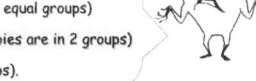

Page 42

1. <u>Multiply</u> and <u>simplify</u> if possible. <u>Color</u> the correct number of images.

Find $\frac{1}{5}$ of 15 flowers.

$\frac{1}{5} \times 15 = \frac{1}{5} \times \frac{15}{1} = \frac{1 \times 15}{5 \times 1} = 3$

Find $\frac{3}{4}$ of 16 stars.

Step 1: $16 \div 4 = 4$ (find $\frac{1}{4}$ out of 16)

Step 2: $3 \times 4 = 12$

Find $\frac{5}{7}$ of 21 flowers.

$\frac{5}{7} \times 21 = \frac{5}{7} \times \frac{21}{1} = \frac{5 \times 21}{7 \times 1} = 15$

Find $\frac{5}{8}$ of 24 stars.

Step 1: $24 \div 8 = 3$ (find $\frac{1}{8}$ out of 24)

Step 2: $5 \times 3 = 15$

Page 43

1. <u>Reduce</u> each fraction to simplest form, <u>write</u> the missing numbers, and <u>compare</u> fractions using "<," ">," or "=." <u>Color</u> the fraction of each shape. <u>Cross out</u> the wrong word.

$\frac{14}{16} = \frac{7}{8} > \frac{14}{20} = \frac{7}{10}$

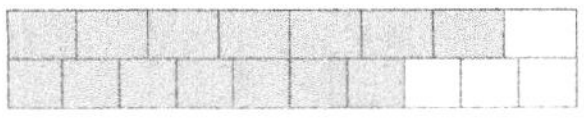

7 out of 8 equal parts are bigger/~~smaller~~ than 7 out of 10 equal parts.

$\frac{16}{18} = \frac{8}{9} > \frac{16}{22} = \frac{8}{11}$

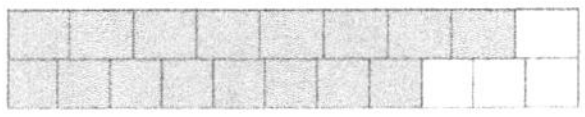

8 out of 9 equal parts are bigger/~~smaller~~ than 8 out of 11 equal parts.

$\frac{20}{24} = \frac{5}{6} > \frac{10}{14} = \frac{5}{7}$

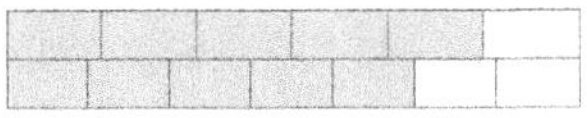

5 out of 6 equal parts are bigger/~~smaller~~ than 5 out of 7 equal parts.

2. <u>Find and circle</u> the area of a rectangle. Hint: Figures are not drawn to scale.

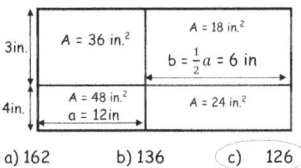

	A = 36 in.²	A = 18 in.²
3in.		$b = \frac{1}{2}a = 6$ in
4in.	A = 48 in.² $a = 12$in	A = 24 in.²

a) 162   b) 136   c) (126)

131

Page 44

1. Write in the missing numbers and cross out the common prime factors.

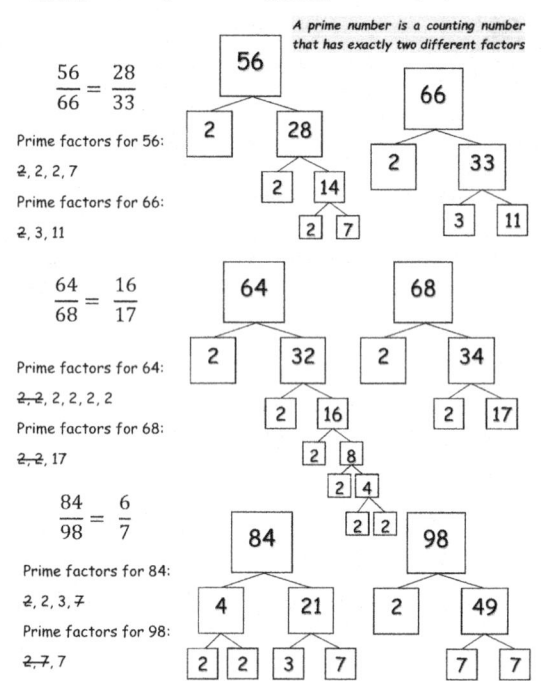

A prime number is a counting number that has exactly two different factors

$\frac{56}{66} = \frac{28}{33}$

Factor tree for 56: 56 → 2, 28 → 2, 14 → 2, 7
Factor tree for 66: 66 → 2, 33 → 3, 11

Prime factors for 56: ~~2~~, 2, 2, 7

Prime factors for 66: ~~2~~, 3, 11

$\frac{64}{68} = \frac{16}{17}$

Factor tree for 64: 64 → 2, 32 → 2, 16 → 2, 8 → 2, 4 → 2, 2
Factor tree for 68: 68 → 2, 34 → 2, 17

Prime factors for 64: ~~2, 2~~, 2, 2, 2, 2

Prime factors for 68: ~~2, 2~~, 17

$\frac{84}{98} = \frac{6}{7}$

Factor tree for 84: 84 → 4, 21 → 2, 2, 3, 7
Factor tree for 98: 98 → 2, 49 → 7, 7

Prime factors for 84: ~~2~~, 2, 3, ~~7~~

Prime factors for 98: ~~2, 7~~, 7

Page 45

1. Reduce each fraction to simplest form, write the missing numbers, and compare fractions using "<," ">," or "=." Color the fraction of each shape. Cross out the wrong word.

$\frac{5}{20} = \frac{1}{4} < \frac{5}{10} = \frac{1}{2}$

1 out of 4 equal parts are ~~bigger~~/smaller than 1 out of 2 equal parts.

$\frac{6}{8} = \frac{3}{4} > \frac{6}{10} = \frac{3}{5}$

3 out of 4 equal parts are bigger/~~smaller~~ than 3 out of 5 equal parts.

$\frac{4}{12} = \frac{1}{3} < \frac{4}{8} = \frac{1}{2}$

1 out of 3 equal parts are ~~bigger~~/smaller than 1 out of 2 equal parts.

$\frac{2}{8} = \frac{1}{4} > \frac{2}{12} = \frac{1}{6}$

1 out of 4 equal parts are bigger/~~smaller~~ than 1 out of 6 equal parts.

2. Write 4 fractions that you cannot reduce to their simplest forms.

$\frac{5}{6}, \frac{3}{7}, \frac{11}{13}, \frac{14}{15}, \frac{2}{5}$

Page 46

1. Write in the missing numbers on the Minecraft factor tree.

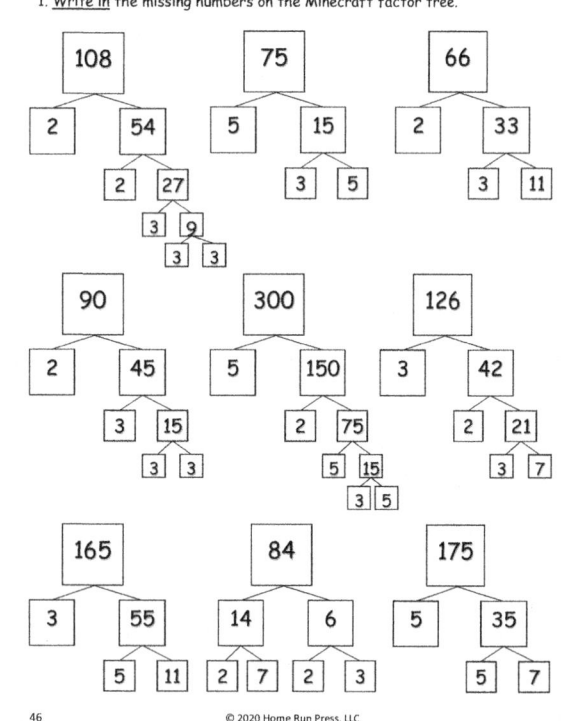

108 → 2, 54 → 2, 27 → 3, 9 → 3, 3

75 → 5, 15 → 3, 5

66 → 2, 33 → 3, 11

90 → 2, 45 → 3, 15 → 3, 3

300 → 5, 150 → 2, 75 → 5, 15 → 3, 5

126 → 3, 42 → 2, 21 → 3, 7

165 → 3, 55 → 5, 11

84 → 14, 6 → 2, 7, 2, 3

175 → 5, 35 → 5, 7

Page 47

1. Write equivalent fractions. Fill in the missing numbers.

$\frac{2 \times 2}{3 \times 2} = \frac{4}{6}$ Common factor = 6 ÷ 3 = 2

$\frac{1 \times 4}{2 \times 4} = \frac{4}{8}$ Common factor = 8 ÷ 2 = 4

$\frac{3 \times 3}{4 \times 3} = \frac{9}{12}$ Common factor = 3

$\frac{1 \times 6}{3 \times 6} = \frac{6}{18}$ Common factor = 6

$\frac{7 \times 3}{8 \times 3} = \frac{21}{24}$ Common factor = 3

$\frac{2 \times 4}{3 \times 4} = \frac{8}{12}$ Common factor = 4

$\frac{5 \times 6}{6 \times 6} = \frac{30}{36}$ Common factor = 6

$\frac{5 \times 4}{7 \times 4} = \frac{20}{28}$ Common factor = 4

$\frac{4 \times 7}{5 \times 7} = \frac{28}{35}$ Common factor = 7

$\frac{2 \times 6}{7 \times 6} = \frac{12}{42}$ Common factor = 6

$\frac{3 \times 12}{5 \times 12} = \frac{36}{60}$ Common factor = 12

$\frac{4 \times 5}{9 \times 5} = \frac{20}{45}$ Common factor = 5

$\frac{2 \times 6}{5 \times 6} = \frac{12}{30}$ Common factor = 6

$\frac{1 \times 6}{8 \times 6} = \frac{6}{48}$ Common factor = 6

$\frac{2 \times 20}{3 \times 20} = \frac{40}{60}$ Common factor = 20

$\frac{1 \times 10}{4 \times 10} = \frac{10}{40}$ Common factor = 10

2. In a florist, $\frac{1}{6}$ of the flowers are daisies and the rest are roses. If there are 30 flowers in all, find the roses.   Answer: $30 - \frac{1}{6} \times 30 = 25$ (roses).

Page 48

1. Multiply and simplify if possible.

Find and color $\frac{2}{3}$ of 21 blocks.

$\frac{2}{3} \times 21 = \frac{2}{3} \times \frac{21}{1} = \frac{2 \times 21}{3 \times 1} = 14$

Find and color $\frac{6}{11}$ of 33 blocks.

Step 1: 33 ÷ 11 = 3 (find $\frac{1}{11}$ out of 33)

Step 2: 6 × 3 = 18

Find and color $\frac{7}{9}$ of 27 blocks.

$\frac{7}{9} \times 27 = \frac{7}{9} \times \frac{27}{1} = \frac{7 \times 27}{9 \times 1} = 21$

Find and color $\frac{4}{7}$ of 28 blocks.

Step 1: 28 ÷ 7 = 4 (find $\frac{1}{7}$ out of 28)

Step 2: 4 × 4 = 16

Page 49

1. Reduce each fraction to simplest form, write the missing numbers, and compare fractions using "<," ">," or "=." Color the fraction of each shape. Cross out the wrong word.

$\frac{36}{48} = \frac{3}{4} > \frac{36}{60} = \frac{3}{5}$

3 out of 4 equal parts are bigger/~~smaller~~ than 3 out of 5 equal parts.

$\frac{45}{70} = \frac{9}{14} > \frac{45}{80} = \frac{9}{16}$

9 out of 14 equal parts are bigger/~~smaller~~ than 9 out of 16 equal parts.

$\frac{40}{64} = \frac{5}{8} > \frac{60}{108} = \frac{5}{9}$

5 out of 8 equal parts are bigger/~~smaller~~ than 5 out of 9 equal parts.

$\frac{28}{49} = \frac{4}{7} > \frac{28}{63} = \frac{4}{9}$

4 out of 7 equal parts are bigger/~~smaller~~ than 4 out of 9 equal parts.

2. Find d and circle the correct answer. Hint: Figures are not drawn to scale.

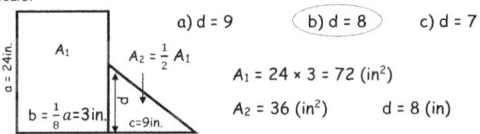

a) d = 9   (b) d = 8)   c) d = 7

$a = 24$ in, A_1, $A_2 = \frac{1}{2} A_1$

$b = \frac{1}{8} a = 3$ in., $c = 9$ in.

$A_1 = 24 \times 3 = 72$ (in^2)
$A_2 = 36$ (in^2)
$d = 8$ (in)

Page 50

1. Multiply and simplify if possible. Circle the right answer (letter).

Find and color $\frac{6}{11}$ of 22 blocks.

$\frac{6}{11} \times 22 = \frac{6}{11} \times \frac{22}{1} = \frac{6 \times 22}{11 \times 1} = 12$

a) 10
(b) 12)
c) 15

Find and color $\frac{7}{10}$ of 30 blocks.

Step 1: 30 ÷ 10 = 3 (find $\frac{1}{10}$ out of 30)

Step 2: 7 × 3 = 21

(a) 21)
b) 35
c) 20

Find and color $\frac{3}{8}$ of 32 blocks.

$\frac{3}{8} \times 32 = \frac{3}{8} \times \frac{32}{1} = \frac{3 \times 32}{8 \times 1} = 12$

a) 16
b) 10
(c) 12)

Find and color $\frac{5}{7}$ of 35 blocks.

Step 1: 35 ÷ 7 = 5 (find $\frac{1}{7}$ out of 35)

Step 2: 5 × 5 = 25

a) 20
b) 35
(c) 25)

Page 51

1. Reduce each fraction to simplest form, write the missing numbers, and compare fractions using "<," ">," or "=." Color the fraction of each shape. Cross out the wrong word.

$\frac{42}{54} = \frac{7}{9} < \frac{42}{48} = \frac{7}{8}$

7 out of 9 equal parts are ~~bigger~~/smaller than 7 out of 8 equal parts.

$\frac{24}{40} = \frac{3}{5} > \frac{24}{64} = \frac{5}{8}$

3 out of 5 equal parts are bigger/~~smaller~~ than 3 out of 8 equal parts.

$\frac{20}{24} = \frac{5}{6} > \frac{40}{64} = \frac{5}{8}$

5 out of 6 equal parts are bigger/~~smaller~~ than 5 out of 8 equal parts.

2. Find and circle the area of a rectangle. Hint: Figures are not drawn to scale.

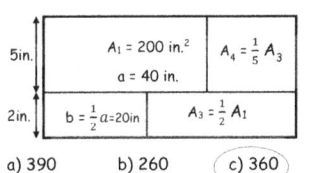

5in., $A_1 = 200$ in.^{2}, $a = 40$ in., $A_4 = \frac{1}{5} A_3$

2in., $b = \frac{1}{2} a = 20$in, $A_3 = \frac{1}{2} A_1$

$A_2 = 2 \times 20 = 40$(in^2)
$A_3 = \frac{1}{2} A_1 = 100$(in^2)
$A_4 = \frac{1}{5} A_3 = 20$(in^2)

a) 390   b) 260   (c) 360)

133

Page 52

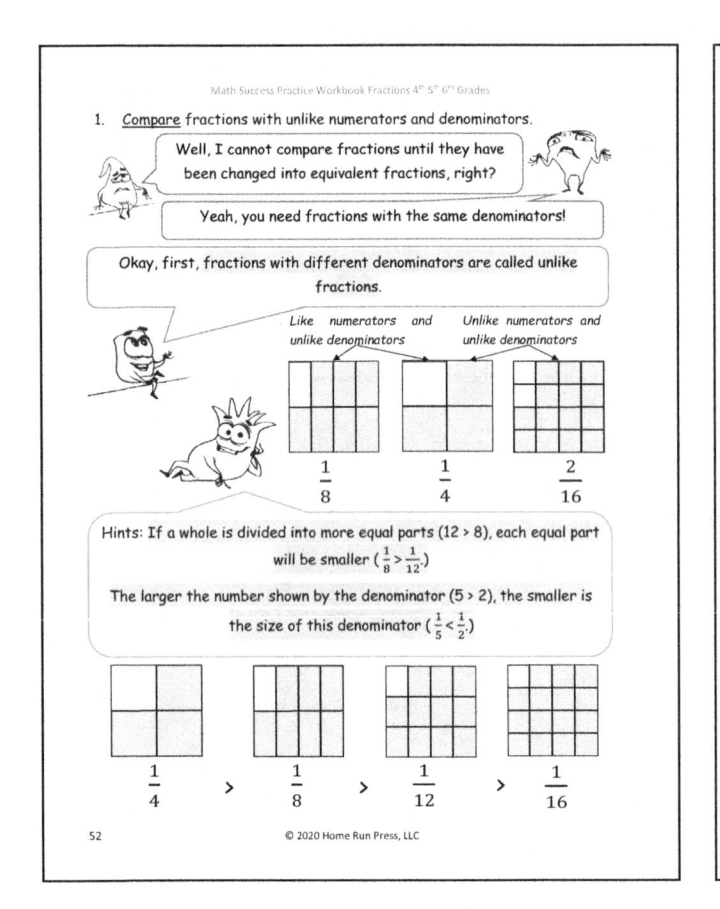

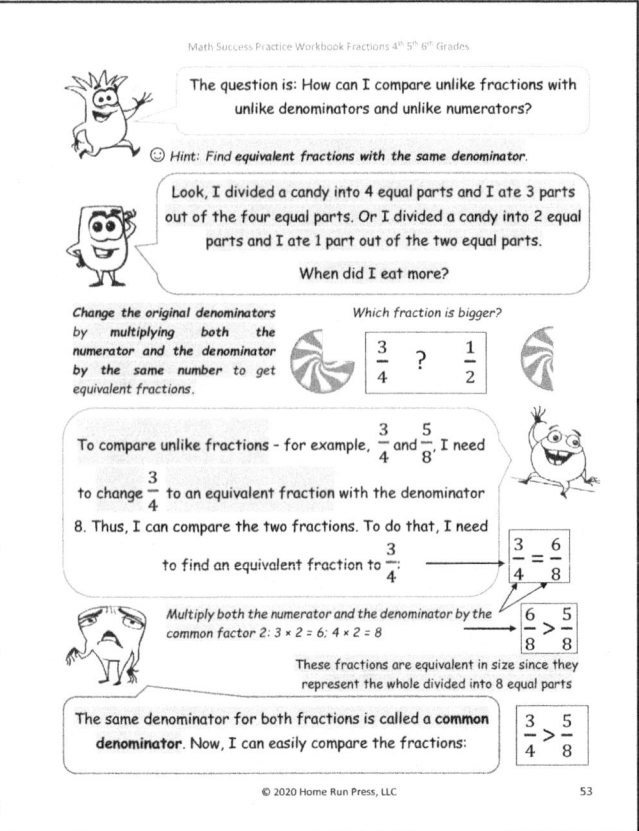

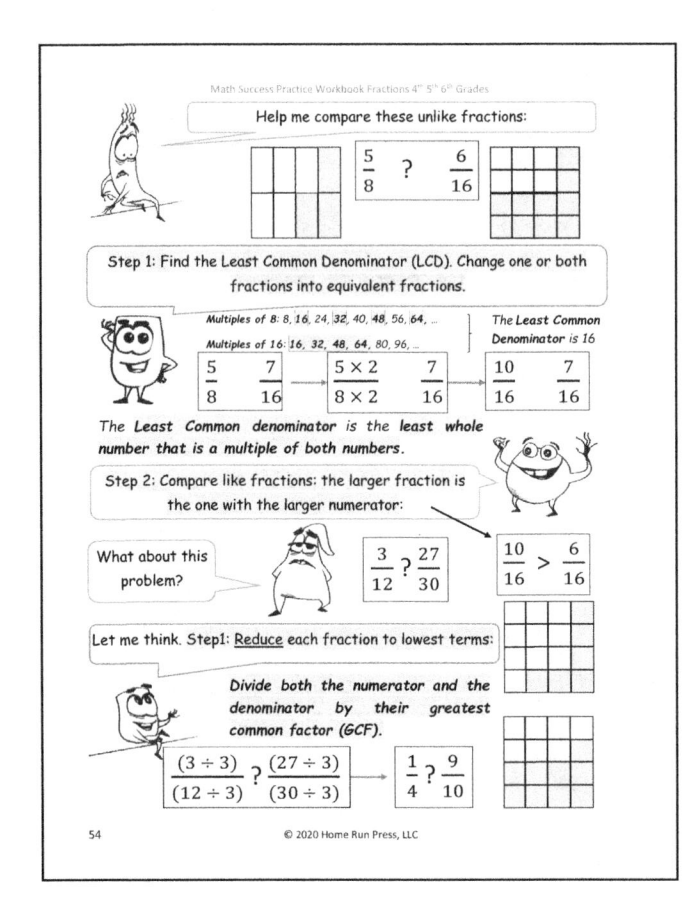

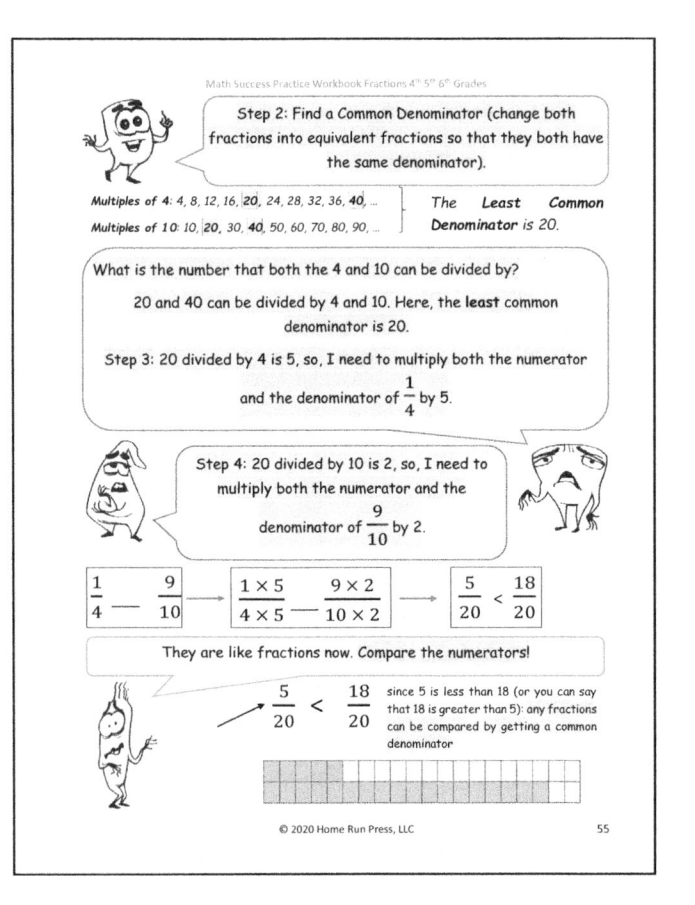

1. <u>Compare</u> fractions with unlike numerators and denominators.

Well, I cannot compare fractions until they have been changed into equivalent fractions, right?

Yeah, you need fractions with the same denominators!

Okay, first, fractions with different denominators are called unlike fractions.

Like numerators and unlike denominators | Unlike numerators and unlike denominators

$\frac{1}{8}$ $\frac{1}{4}$ $\frac{2}{16}$

Hints: If a whole is divided into more equal parts (12 > 8), each equal part will be smaller ($\frac{1}{8} > \frac{1}{12}$.)

The larger the number shown by the denominator (5 > 2), the smaller is the size of this denominator ($\frac{1}{5} < \frac{1}{2}$.)

$\frac{1}{4} > \frac{1}{8} > \frac{1}{12} > \frac{1}{16}$

Page 53

The question is: How can I compare unlike fractions with unlike denominators and unlike numerators?

☺ Hint: Find *equivalent fractions with the same denominator.*

Look, I divided a candy into 4 equal parts and I ate 3 parts out of the four equal parts. Or I divided a candy into 2 equal parts and I ate 1 part out of the two equal parts.

When did I eat more?

Change the original denominators by multiplying both the numerator and the denominator by the same number to get equivalent fractions.

Which fraction is bigger? $\frac{3}{4}$? $\frac{1}{2}$

To compare unlike fractions - for example, $\frac{3}{4}$ and $\frac{5}{8}$, I need to change $\frac{3}{4}$ to an equivalent fraction with the denominator 8. Thus, I can compare the two fractions. To do that, I need to find an equivalent fraction to $\frac{3}{4}$:

$\frac{3}{4} = \frac{6}{8}$

Multiply both the numerator and the denominator by the common factor 2: $3 \times 2 = 6$; $4 \times 2 = 8$

$\frac{6}{8} > \frac{5}{8}$

These fractions are equivalent in size since they represent the whole divided into 8 equal parts

The same denominator for both fractions is called a **common denominator**. Now, I can easily compare the fractions: $\frac{3}{4} > \frac{5}{8}$

Page 54

Help me compare these unlike fractions:

$\frac{5}{8}$? $\frac{6}{16}$

Step 1: Find the Least Common Denominator (LCD). Change one or both fractions into equivalent fractions.

Multiples of 8: 8, 16, 24, 32, 40, 48, 56, 64, ...
Multiples of 16: 16, 32, 48, 64, 80, 96, ...

The Least Common Denominator is 16

$\frac{5}{8}$ $\frac{7}{16}$ → $\frac{5 \times 2}{8 \times 2}$ $\frac{7}{16}$ → $\frac{10}{16}$ $\frac{7}{16}$

The Least Common denominator is the least whole number that is a multiple of both numbers.

Step 2: Compare like fractions: the larger fraction is the one with the larger numerator:

$\frac{10}{16} > \frac{6}{16}$

What about this problem? $\frac{3}{12}$? $\frac{27}{30}$

Let me think. Step 1: <u>Reduce</u> each fraction to lowest terms:

Divide both the numerator and the denominator by their greatest common factor (GCF).

$\frac{(3 \div 3)}{(12 \div 3)}$? $\frac{(27 \div 3)}{(30 \div 3)}$ → $\frac{1}{4}$? $\frac{9}{10}$

Page 55

Step 2: Find a Common Denominator (change both fractions into equivalent fractions so that they both have the same denominator).

Multiples of 4: 4, 8, 12, 16, **20**, 24, 28, 32, 36, **40**, ...
Multiples of 10: 10, **20**, 30, **40**, 50, 60, 70, 80, 90, ...

The Least Common Denominator is 20.

What is the number that both the 4 and 10 can be divided by?

20 and 40 can be divided by 4 and 10. Here, the **least** common denominator is 20.

Step 3: 20 divided by 4 is 5, so, I need to multiply both the numerator and the denominator of $\frac{1}{4}$ by 5.

Step 4: 20 divided by 10 is 2, so, I need to multiply both the numerator and the denominator of $\frac{9}{10}$ by 2.

$\frac{1}{4}$ $\frac{9}{10}$ → $\frac{1 \times 5}{4 \times 5}$ $\frac{9 \times 2}{10 \times 2}$ → $\frac{5}{20} < \frac{18}{20}$

They are like fractions now. Compare the numerators!

$\frac{5}{20} < \frac{18}{20}$ since 5 is less than 18 (or you can say that 18 is greater than 5): any fractions can be compared by getting a common denominator

Page 56

1. <u>Color</u> the correct fraction of each shape. <u>Write</u> the missing numbers. <u>Compare</u> unlike fractions using ">," "<," or "=."

$\dfrac{1}{6} < \dfrac{2}{4}$ The Least Common Denominator is 12: $\dfrac{1 \times 2}{6 \times 2} < \dfrac{2 \times 3}{4 \times 3}$ → $\dfrac{2}{12} < \dfrac{6}{12}$

Multiples of 6: 6, 12, 18, 24, …
Multiples of 4: 4, 8, 12, 16, 20

$\dfrac{3}{5} < \dfrac{2}{3}$ The Least Common Denominator is 15: $\dfrac{3 \times 3}{5 \times 3} < \dfrac{2 \times 5}{3 \times 5}$ → $\dfrac{9}{15} < \dfrac{10}{15}$

Multiples of 5: 5, 10, 15
Multiples of 3: 3, 6, 9, 12, 15

$\dfrac{4}{8} < \dfrac{5}{6}$ The Least Common Denominator is 24: $\dfrac{4 \times 3}{8 \times 3} < \dfrac{5 \times 4}{6 \times 4}$ → $\dfrac{12}{24} < \dfrac{20}{24}$

Multiples of 8: 8, 16, 24
Multiples of 6: 6, 12, 18, 24

$\dfrac{3}{4} < \dfrac{5}{6}$ The Least Common Denominator is 12: $\dfrac{3 \times 3}{4 \times 3} < \dfrac{5 \times 2}{6 \times 2}$ → $\dfrac{9}{12} < \dfrac{10}{12}$

Multiples of 4: 4, 8, 12
Multiples of 6: 6, 12

Page 57

1. <u>Color</u> the correct fraction of each shape. <u>Write</u> the missing numbers. <u>Compare</u> unlike fractions using ">," "<," or "=."

$\dfrac{5}{6} > \dfrac{7}{9}$ The Least Common Denominator is 18: $\dfrac{5 \times 3}{6 \times 3} > \dfrac{7 \times 2}{9 \times 2}$ → $\dfrac{15}{18} > \dfrac{14}{18}$

Multiples of 6: 6, 12, 18
Multiples of 9: 9, 18

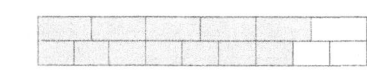

$\dfrac{2}{3} \;—\; \dfrac{9}{10}$ The Least Common Denominator is 30: $\dfrac{2 \times 10}{3 \times 10} < \dfrac{9 \times 3}{10 \times 3}$ → $\dfrac{20}{30} < \dfrac{27}{30}$

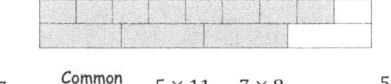

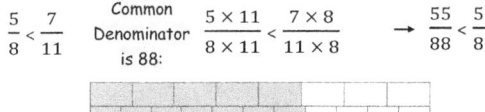

Multiples of 3: 3, 6, 9, 12, 15, 18, 21, 24, 27, 30
Multiples of 10: 10, 20, 30

2. <u>Multiply</u> and <u>simplify</u> if possible. <u>Circle</u> the correct answer (letter). <u>Find</u> and <u>color red</u> $\dfrac{4}{5}$ of 20 blocks. <u>Find</u> and <u>cross out</u> $\dfrac{3}{4}$ of these red blocks.

$\dfrac{4}{5} \times 20 = \dfrac{4}{5} \times \dfrac{20}{1} = \dfrac{4 \times 20}{5 \times 1} = 16$

$\dfrac{3}{4} \times 16 = \dfrac{3}{4} \times \dfrac{16}{1} = \dfrac{3 \times 16}{4 \times 1} = 12$

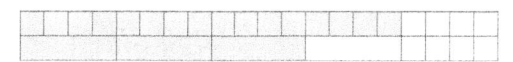

a) 12 (circled)
b) 16
c) 10

Page 58

1. <u>Color</u> the correct fraction of each shape. <u>Write</u> the missing numbers. <u>Compare</u> unlike fractions using ">," "<," or "=."

$\dfrac{8}{9} > \dfrac{3}{4}$ Common Denominator is 36: $\dfrac{8 \times 4}{9 \times 4} > \dfrac{3 \times 9}{4 \times 9}$ → $\dfrac{32}{36} > \dfrac{27}{36}$

$\dfrac{5}{8} < \dfrac{7}{11}$ Common Denominator is 88: $\dfrac{5 \times 11}{8 \times 11} < \dfrac{7 \times 8}{11 \times 8}$ → $\dfrac{55}{88} < \dfrac{56}{88}$

2. <u>Multiply</u> and <u>simplify</u> if possible. <u>Circle</u> the correct answer (letter). <u>Find</u> and <u>color red</u> $\dfrac{4}{7}$ of 21 blocks. <u>Find</u> and <u>cross out</u> $\dfrac{1}{2}$ of these red blocks.

$\dfrac{4}{7} \times 21 = \dfrac{4}{7} \times \dfrac{21}{1} = \dfrac{4 \times 21}{7 \times 1} = 12$

$\dfrac{1}{2} \times 12 = \dfrac{1}{2} \times \dfrac{12}{1} = \dfrac{1 \times 12}{2 \times 1} = 6$

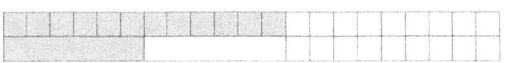

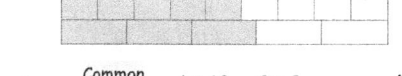

a) 9
b) 10
c) 6 (circled)

Page 59

1. <u>Color</u> the correct fraction of each shape. <u>Write</u> the missing numbers. <u>Compare</u> unlike fractions using ">," "<," or "=."

$\dfrac{5}{9} < \dfrac{3}{5}$ Common Denominator is 45: $\dfrac{5 \times 5}{9 \times 5} < \dfrac{3 \times 9}{5 \times 9}$ → $\dfrac{25}{45} < \dfrac{27}{45}$

$\dfrac{4}{9} < \dfrac{6}{10}$ Common Denominator is 90: $\dfrac{4 \times 10}{9 \times 10} < \dfrac{6 \times 9}{10 \times 9}$ → $\dfrac{40}{90} < \dfrac{54}{90}$

2. <u>One fifth</u> of a bottle can hold 200 ml of water. <u>How much water</u> will the bottle hold when it is $\dfrac{1}{2}$ full? <u>Circle</u> the correct answer.

a) 400   b) 1000   c) 750   d) 500 (circled)

3. What fraction is made of flowers? $\dfrac{8}{12}$

Page 60

Math Success Practice Workbook Fractions 4th 5th 6th Grades

1. <u>Color</u> the correct fraction of each shape. <u>Write</u> the missing numbers. <u>Compare</u> unlike fractions using ">," "<," or "=."

Simplify and factor 2 out of 6 and 14 to avoid greater numbers

$\frac{6}{14} < \frac{5}{9}$ Common Denominator is 63: $\frac{3 \times 9}{7 \times 9} < \frac{5 \times 7}{9 \times 7}$ → $\frac{27}{63} < \frac{35}{63}$

$\frac{1}{3} < \frac{3}{7}$ Common Denominator is 21: $\frac{1 \times 7}{3 \times 7} < \frac{3 \times 3}{7 \times 3}$ → $\frac{7}{21} < \frac{9}{21}$

2. It takes $2\frac{1}{3}$ hours to paint a room. <u>How many hours</u> will it take to paint 3 such rooms? <u>Circle</u> the correct answer.

a) 4   b) 8   c) 6   (d) 7

3. What fraction is made of stars?

$\frac{4}{12} = \frac{1}{3}$

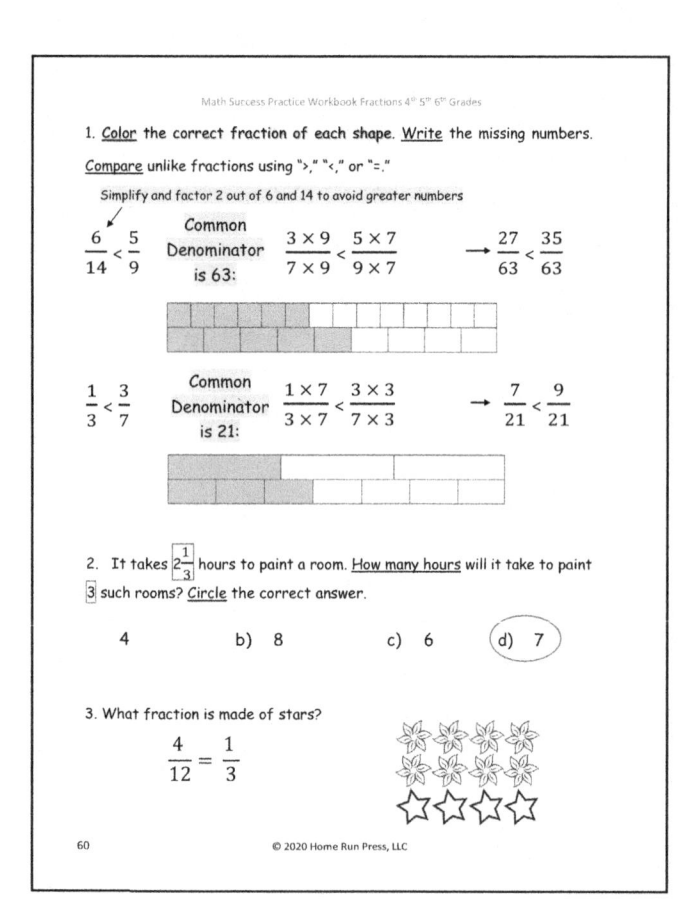

Page 61

1. <u>Color</u> the numerator. <u>Find</u> the Common Denominator. <u>Compare</u> unlike fractions.

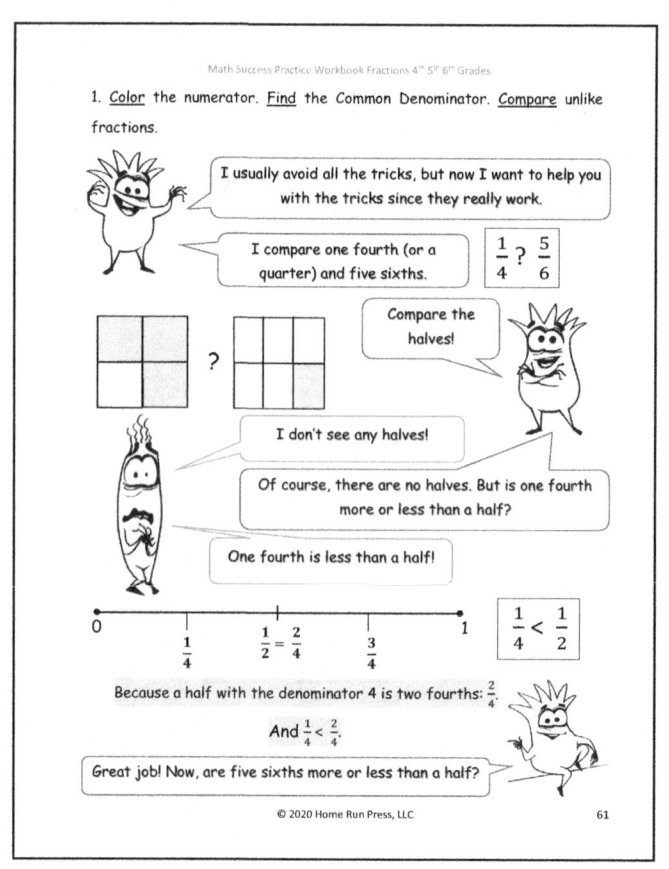

I usually avoid all the tricks, but now I want to help you with the tricks since they really work.

I compare one fourth (or a quarter) and five sixths. $\frac{1}{4} ? \frac{5}{6}$

Compare the halves!

I don't see any halves!

Of course, there are no halves. But is one fourth more or less than a half?

One fourth is less than a half!

0 — $\frac{1}{4}$ — $\frac{1}{2} = \frac{2}{4}$ — $\frac{3}{4}$ — 1 $\frac{1}{4} < \frac{1}{2}$

Because a half with the denominator 4 is two fourths: $\frac{2}{4}$.

And $\frac{1}{4} < \frac{2}{4}$.

Great job! Now, are five sixths more or less than a half?

Page 62

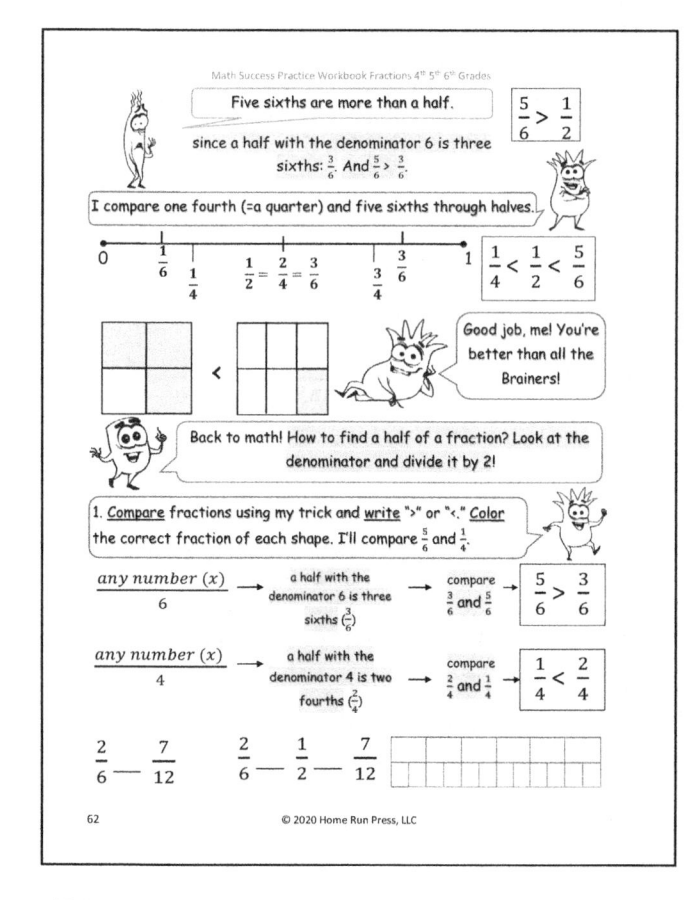

Five sixths are more than a half. $\frac{5}{6} > \frac{1}{2}$

since a half with the denominator 6 is three sixths: $\frac{3}{6}$. And $\frac{5}{6} > \frac{3}{6}$.

I compare one fourth (=a quarter) and five sixths through halves.

0 — $\frac{1}{6}$ $\frac{1}{4}$ — $\frac{1}{2} = \frac{2}{4} = \frac{3}{6}$ — $\frac{3}{4}$ $\frac{5}{6}$ — 1 $\frac{1}{4} < \frac{1}{2} < \frac{5}{6}$

Good job, me! You're better than all the Brainers!

Back to math! How to find a half of a fraction? Look at the denominator and divide it by 2!

1. <u>Compare</u> fractions using my trick and <u>write</u> ">" or "<." <u>Color</u> the correct fraction of each shape. I'll compare $\frac{5}{6}$ and $\frac{1}{4}$.

$\frac{any\ number\ (x)}{6}$ → a half with the denominator 6 is three sixths ($\frac{3}{6}$) → compare $\frac{3}{6}$ and $\frac{5}{6}$ → $\frac{5}{6} > \frac{3}{6}$

$\frac{any\ number\ (x)}{4}$ → a half with the denominator 4 is two fourths ($\frac{2}{4}$) → compare $\frac{2}{4}$ and $\frac{1}{4}$ → $\frac{1}{4} < \frac{2}{4}$

$\frac{2}{6}$ — $\frac{7}{12}$ $\frac{2}{6}$ — $\frac{1}{2}$ — $\frac{7}{12}$

Page 63

1. <u>Compare</u> fractions using my trick and <u>write</u> ">," "=," or "<." <u>Color</u> the correct fraction of each shape.

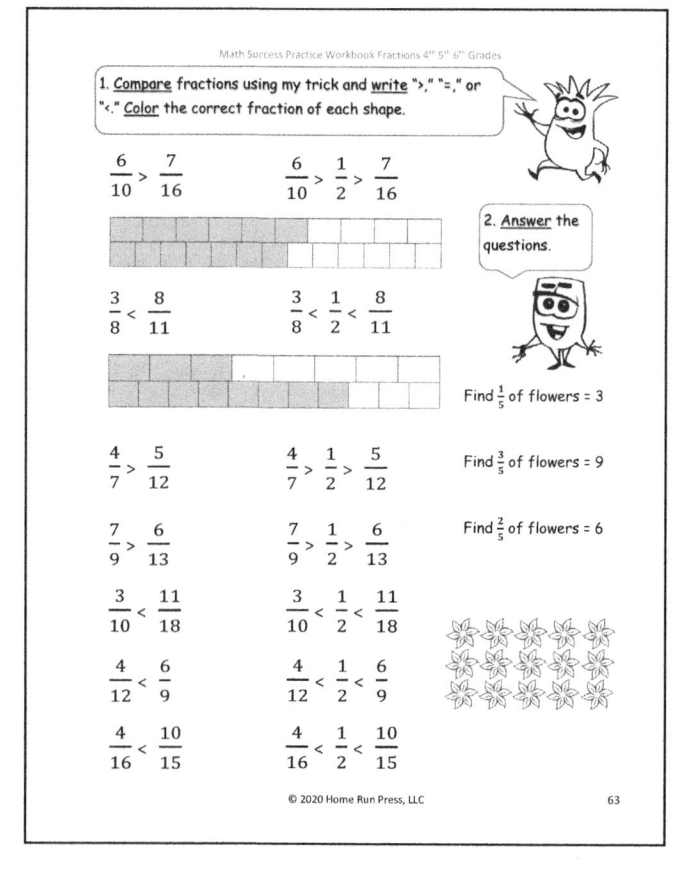

$\frac{6}{10} > \frac{7}{16}$ $\frac{6}{10} > \frac{1}{2} > \frac{7}{16}$

2. <u>Answer</u> the questions.

$\frac{3}{8} < \frac{8}{11}$ $\frac{3}{8} < \frac{1}{2} < \frac{8}{11}$

Find $\frac{1}{5}$ of flowers = 3

$\frac{4}{7} > \frac{5}{12}$ $\frac{4}{7} > \frac{1}{2} > \frac{5}{12}$

Find $\frac{3}{5}$ of flowers = 9

$\frac{7}{9} > \frac{6}{13}$ $\frac{7}{9} > \frac{1}{2} > \frac{6}{13}$

Find $\frac{2}{5}$ of flowers = 6

$\frac{3}{10} < \frac{11}{18}$ $\frac{3}{10} < \frac{1}{2} < \frac{11}{18}$

$\frac{4}{12} < \frac{6}{9}$ $\frac{4}{12} < \frac{1}{2} < \frac{6}{9}$

$\frac{4}{16} < \frac{10}{15}$ $\frac{4}{16} < \frac{1}{2} < \frac{10}{15}$

Page 64

What's the difference between a proper fraction and an improper fraction?

That's easy! An improper fraction always misbehaves☺! Just like me!

A proper fraction is less than a whole: that means the numerator (top number) is less than the denominator (bottom number):

a whole > $\frac{1}{2}$ (a half);

a whole > $\frac{3}{4}$ (three fourths).

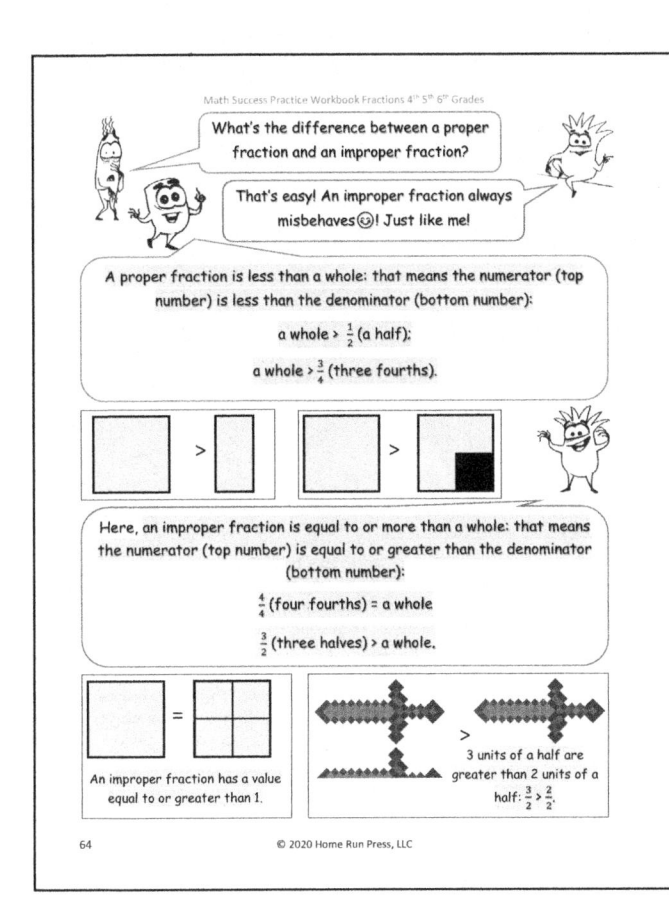

Here, an improper fraction is equal to or more than a whole: that means the numerator (top number) is equal to or greater than the denominator (bottom number):

$\frac{4}{4}$ (four fourths) = a whole

$\frac{3}{2}$ (three halves) > a whole.

An improper fraction has a value equal to or greater than 1.

3 units of a half are greater than 2 units of a half: $\frac{3}{2} > \frac{2}{2}$.

Page 65

I would say that we compare 1 whole plus $\frac{1}{2}$ and 1 whole.

Numbers like 1, 2, 3, 4, etc. are called whole numbers

You mentioned a mixed fractions. A combination of a whole number with a fraction is called a mixed number. $1\frac{1}{2}$ or $3\frac{2}{5}$ are called mixed numbers.

Hint: A mixed number = the sum of a whole number and a fraction: $1\frac{1}{2} = 1 + \frac{1}{2}$

1. <u>Cross out</u> proper fractions and <u>circle</u> improper fractions. <u>Simplify</u> if possible. Be quick! I need to build a house!

$\frac{8}{3} = 2\frac{2}{3}$ $\frac{5}{9} = —$ $\frac{4}{8} = \frac{1}{2}$ $\frac{6}{2} = 3$

$\frac{1}{12} = —$ $\frac{4}{3} = 1\frac{1}{3}$ $\frac{15}{16} = —$ $\frac{20}{11} = 1\frac{9}{11}$

$\frac{7}{10} = —$ $\frac{3}{8} = —$ $\frac{21}{16} = 1\frac{5}{16}$ $\frac{18}{9} = 2$

$\frac{3}{9} = \frac{1}{3}$ $\frac{14}{3} = 4\frac{2}{3}$ $\frac{1}{5} = —$ $\frac{3}{6} = \frac{1}{2}$

$\frac{7}{5} = 1\frac{2}{5}$ $\frac{6}{8} = \frac{3}{4}$ $\frac{3}{24} = \frac{1}{8}$ $\frac{36}{12} = 3$

Page 66

1. <u>Draw</u> the swords to match each fraction. <u>Circle</u> the proper fractions.

$\frac{4}{2}$

$\frac{2}{2}$

2. <u>Color</u> the blocks black to match each fraction. <u>Circle</u> the improper fraction(s). <u>Cross out</u> the block(s) you do not need.

$\frac{1}{4}$

$\frac{6}{4}$

$\frac{8}{16}$

3. <u>Four-sevenths</u> of $\boxed{42}$ villagers are harvesting crops, and $\boxed{7}$ villagers are planting seeds. <u>How many villagers</u> are picking up carrots?

$\frac{4}{7} \times 42 = 24$ (crops)   42 - 24 - 7 = 11 (carrots)

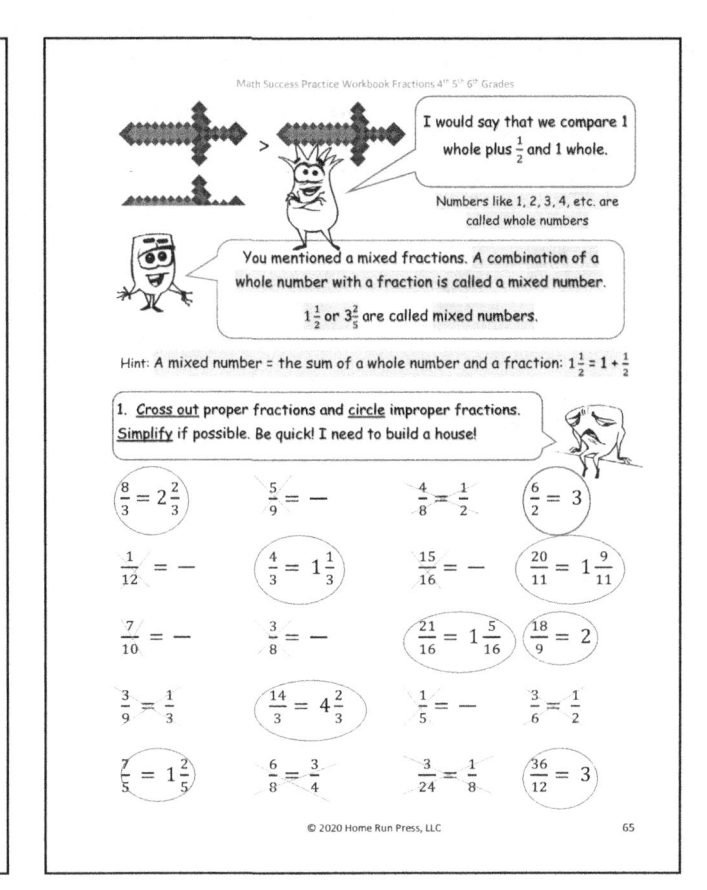

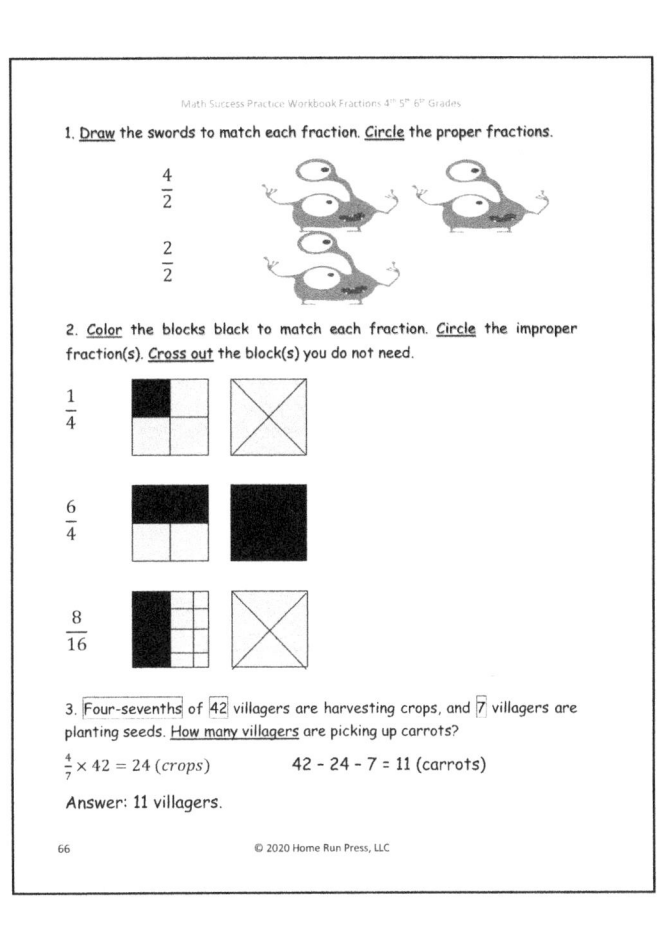

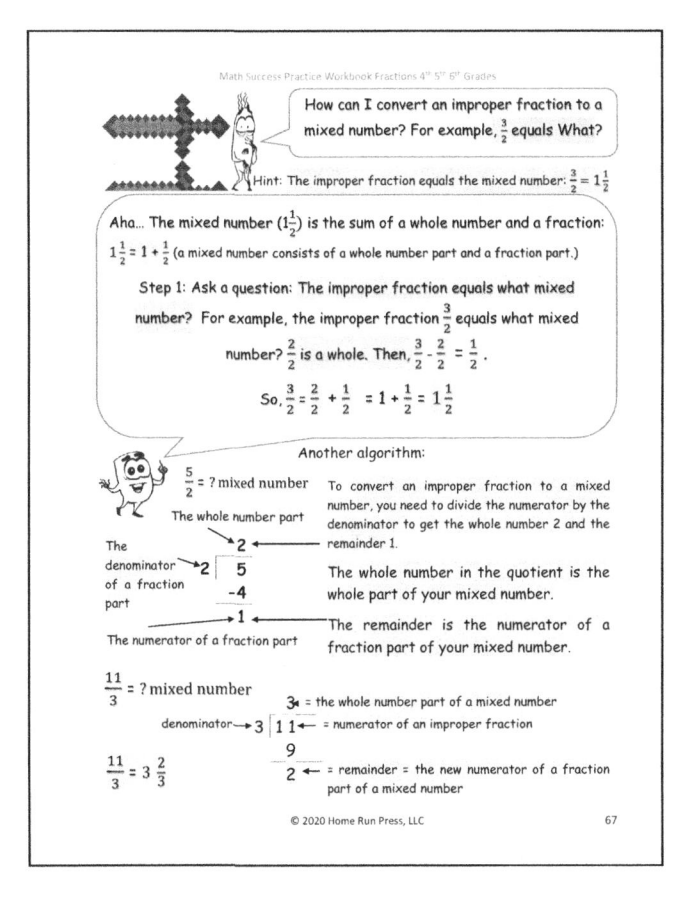

Answer: 11 villagers.

Page 67

How can I convert an improper fraction to a mixed number? For example, $\frac{3}{2}$ equals What?

Hint: The improper fraction equals the mixed number: $\frac{3}{2} = 1\frac{1}{2}$

Aha... The mixed number ($1\frac{1}{2}$) is the sum of a whole number and a fraction: $1\frac{1}{2} = 1 + \frac{1}{2}$ (a mixed number consists of a whole number part and a fraction part.)

Step 1: Ask a question: The improper fraction equals what mixed number? For example, the improper fraction $\frac{3}{2}$ equals what mixed number? $\frac{2}{2}$ is a whole. Then, $\frac{3}{2} - \frac{2}{2} = \frac{1}{2}$.

So, $\frac{3}{2} = \frac{2}{2} + \frac{1}{2} = 1 + \frac{1}{2} = 1\frac{1}{2}$

Another algorithm:

$\frac{5}{2}$ = ? mixed number

To convert an improper fraction to a mixed number, you need to divide the numerator by the denominator to get the whole number 2 and the remainder 1.

```
       2
    2 ) 5
       -4
        1
```

The whole number in the quotient is the whole part of your mixed number.

The remainder is the numerator of a fraction part of your mixed number.

$\frac{11}{3}$ = ? mixed number

```
          3  ← the whole number part of a mixed number
denominator → 3 ) 1 1 ← numerator of an improper fraction
              9
              2 = remainder = the new numerator of a fraction part of a mixed number
```

$\frac{11}{3} = 3\frac{2}{3}$

137

Page 68

1. <u>Convert</u> an improper fraction to a mixed number. <u>Color</u> the blocks to match each fraction.

$\frac{11}{5} = 2\frac{1}{5}$ $\frac{19}{8} = 2\frac{3}{8}$ $\frac{13}{4} = 3\frac{1}{4}$ $\frac{15}{2} = 7\frac{1}{2}$

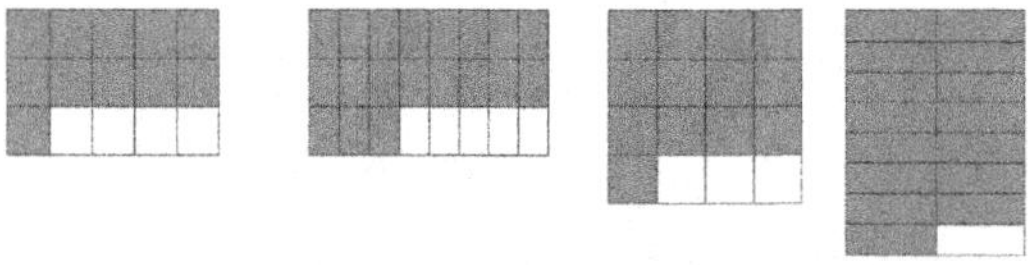

2. Two-eighths of 72 kids are boys, and four-ninths of the rest of the kids are girls. How many kids are girls?

$\frac{2}{8} \times 72 = 18$ (boys) $\frac{4}{9} \times 54 = 24$ (girls)

Answer: 24.

3. <u>Convert</u> an improper fraction to a mixed number.

$\frac{8}{3} = 2\frac{2}{3}$ $\frac{9}{5} = 1\frac{4}{5}$ $\frac{8}{4} = 2$ $\frac{6}{2} = 3$

$\frac{12}{5} = 2\frac{2}{5}$ $\frac{10}{3} = 3\frac{1}{3}$ $\frac{16}{8} = 2$ $\frac{20}{11} = 1\frac{9}{11}$

$\frac{18}{10} = 1\frac{4}{5}$ $\frac{24}{7} = 3\frac{3}{7}$ $\frac{20}{16} = 1\frac{1}{4}$ $\frac{18}{9} = 2$

Page 69

1. <u>Convert</u> an improper fraction to a mixed number. <u>Simplify</u>.

$\frac{9}{3} = 3$ $\frac{14}{3} = 4\frac{2}{3}$ $\frac{5}{2} = 2\frac{1}{2}$ $\frac{7}{6} = 1\frac{1}{6}$

$\frac{7}{5} = 1\frac{2}{5}$ $\frac{8}{5} = 1\frac{3}{5}$ $\frac{24}{3} = 8$ $\frac{36}{12} = 3$

$\frac{22}{3} = 7\frac{1}{3}$ $\frac{34}{6} = 5\frac{2}{3}$ $\frac{11}{2} = 5\frac{1}{2}$ $\frac{28}{5} = 5\frac{3}{5}$

$\frac{42}{9} = 4\frac{2}{3}$ $\frac{25}{4} = 6\frac{1}{4}$ $\frac{63}{8} = 7\frac{7}{8}$ $\frac{36}{10} = 2\frac{3}{5}$

$\frac{31}{7} = 4\frac{3}{7}$ $\frac{46}{5} = 9\frac{1}{5}$ $\frac{25}{12} = 2\frac{1}{12}$ $\frac{41}{15} = 2\frac{11}{15}$

2. Two-sevenths of 35 fruits are apples and the rest are oranges. How many more or less apples than oranges are there?

$\frac{2}{7} \times 35 = 10$ (apples) $35 - 10 = 25$ (oranges)

Answer: 15 less.

3. I have run away from five-eighths of the 32 zombies I had met in my game. My brother run away from four-ninths of the 27 zombies he had met in his game. Who was luckier and run away from the greatest number of zombies?

$\frac{5}{8} \times 32 = 20$ (I) $\frac{4}{9} \times 27 = 12$ (Brother)

Answer: I was luckier.

Page 70

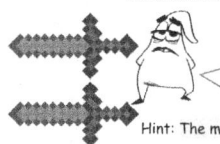

How can I convert a mixed number to an improper fraction?
For example, $2\frac{2}{3}$ equals What?

Hint: The mixed number equals the improper fraction:
$2\frac{1}{2}$ (two swords and a half) = $\frac{5}{2}$ (five halves of swords)

Correct. Let's think!
Step 1: The whole number part equals what fraction? The whole number part of 1 equals: $1 = \frac{2}{2}$. The whole number part of 2 equals: $2 = \frac{4}{2}$.
Step 2: Add the two fractions: $\frac{4}{2} + \frac{1}{2} = \frac{4+1}{2} = \frac{5}{2}$

Another algorithm: $2\frac{1}{2} = ?$

Step 1: Multiply the whole number part by the denominator (in our problem we multiply 2 by 2). → $2 \times 2 = 4$ (the product is the numerator of the whole number part)

Step 2: Add the product to the numerator of a fraction part (in our problem it's 1). → $4 + 1 = 5$ (we added the numerator of the whole number part and the numerator of the fraction part that equals the numerator of the improper fraction)

$2\frac{1}{2} = \frac{5}{2}$

Page 71

1. <u>Convert</u> a mixed number to an improper fraction.

$1\frac{3}{5} = \frac{8}{5}$ $2\frac{5}{8} = \frac{21}{8}$ $3\frac{1}{4} = \frac{13}{8}$

1) $1 \times 5 = 5$ 1) $2 \times 8 = 16$ 1) $3 \times 4 = 12$
2) $5 + 3 = 8$ 2) $16 + 5 = 21$ 2) $12 + 1 = 13$

2. Three-tenths of 50 blooming flowers were blue orchid. If sunflower were one-tenths more than blue orchid, and the rest of the flowers were divided by peonies and lilacs, and roses respectively, how many lilacs were blooming?

$\frac{3}{10} \times 50 = 15$ (BO) $\frac{4}{10} \times 50 = 20$ (S)

Answer: $\frac{1}{10} \times 50 = 5$ (lilacs)

3. <u>Convert</u> a mixed number to an improper fraction. <u>Write</u> the missing numbers.

$4\frac{2}{5} = \frac{(4 \times 5)+2}{5} = \frac{22}{5}$ $7\frac{1}{2} = \frac{(7 \times 2)+1}{2} = \frac{15}{2}$

$2\frac{1}{4} = \frac{(2 \times 4)+1}{4} = \frac{9}{4}$ $6\frac{2}{3} = \frac{(6 \times 3)+2}{3} = \frac{20}{3}$

$3\frac{1}{2} = \frac{(3 \times 2)+1}{2} = \frac{7}{2}$ $3\frac{5}{8} = \frac{(3 \times 8)+5}{8} = \frac{29}{8}$

Page 72

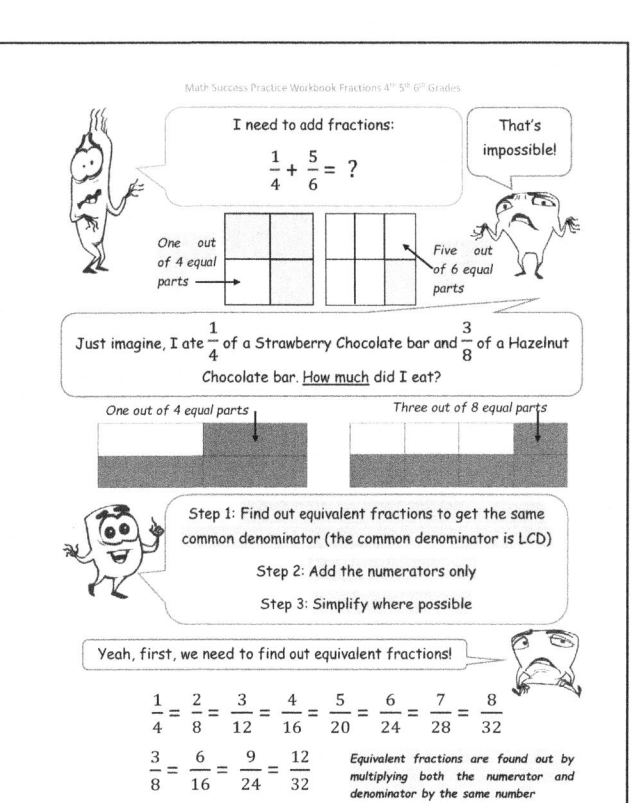

1. <u>Convert</u> a mixed number to an improper fraction. <u>Write</u> the missing numbers.

$7\frac{3}{4} = \frac{(7\times4)+3}{4} = \frac{31}{4}$ $9\frac{4}{5} = \frac{(9\times5)+4}{5} = \frac{49}{5}$

$3\frac{3}{8} = \frac{(3\times8)+3}{8} = \frac{27}{8}$ $5\frac{5}{8} = \frac{(5\times8)+5}{8} = \frac{45}{8}$

$6\frac{1}{10} = \frac{(6\times10)+1}{10} = \frac{61}{10}$ $3\frac{2}{9} = \frac{(3\times9)+2}{9} = \frac{29}{9}$

$10\frac{3}{5} = \frac{(10\times5)+3}{5} = \frac{53}{5}$ $7\frac{9}{11} = \frac{(7\times11)+9}{11} = \frac{86}{11}$

2. I met some horses of two colors. Three-sixths of the horses were white. 24 horses were black. How many horses did I meet?

$\frac{3}{6} \times x = white$ $\frac{3}{6} = 24 = black$ $\frac{6}{6} = white + black = 48$

Answer: 48

3. The villager planted some melon and pumpkin seeds. Two-fifths of them were melon seeds. If he planted 45 pumpkin seeds, how many melon and pumpkin seeds did he have at first?

$\frac{3}{5} \times x = 45$ → $\frac{1}{5} = 15$ → $\frac{5}{5} = 75$

Answer: 75 seeds.

Page 73

1. <u>Convert</u> an improper fraction to a mixed number. <u>Circle</u> the correct answer.

$\frac{9}{2} = 4\frac{1}{2}$ $\frac{25}{8} = 3\frac{1}{8}$ $\frac{31}{4} = 7\frac{3}{4}$ $\frac{37}{5} = 7\frac{2}{5}$

a) $5\frac{2}{2}$ a) $2\frac{5}{8}$ a) $5\frac{3}{4}$ (a) $7\frac{2}{5}$
b) $1\frac{3}{2}$ (b) $3\frac{1}{8}$ b) $8\frac{1}{4}$ b) $6\frac{7}{5}$
(c) $4\frac{1}{2}$ c) $5\frac{1}{8}$ c) $4\frac{3}{4}$ c) $5\frac{21}{5}$
d) $4\frac{2}{2}$ d) $3\frac{2}{8}$ (d) $7\frac{3}{4}$ d) $4\frac{4}{5}$

2. I am an improper fraction. The sum of my numerator and denominator is 14, their difference is 2^3 (2 × 2 × 2). What fraction am I?

a) $\frac{8}{6}$ b) $\frac{15}{1}$ c) $\frac{6}{8}$ (d) $\frac{11}{3}$

3. <u>True or false?</u> <u>Explain</u>. <u>Circle</u> the correct answer.

To convert an improper fraction to a mixed number, you need to divide the denominator by the numerator.

To convert an improper fraction to a mixed number, you need to divide the numerator by the denominator.

a)   TRUE   (b) FALSE

Page 74

1. <u>Convert</u> a mixed number to an improper fraction. <u>Circle</u> the correct answer.

$2\frac{1}{4} = \frac{9}{4}$ $7\frac{3}{8} = \frac{59}{8}$ $5\frac{7}{12} = \frac{67}{12}$ $3\frac{2}{5} = \frac{17}{5}$

a) $\frac{8}{4}$ a) $\frac{60}{8}$ a) $\frac{66}{12}$ (a) $\frac{17}{5}$
b) $\frac{3}{4}$ b) $\frac{56}{8}$ b) $\frac{70}{12}$ b) $\frac{10}{5}$
(c) $\frac{9}{4}$ c) $\frac{58}{8}$ (c) $\frac{67}{12}$ c) $\frac{21}{5}$
d) $\frac{10}{4}$ (d) $\frac{59}{8}$ d) $\frac{77}{12}$ d) $\frac{16}{5}$

2. I am an improper fraction. The product of my numerator and denominator is 27, their quotient is 3. What fraction am I?

(a) $\frac{27}{9}$ b) $\frac{15}{14}$ c) $\frac{12}{4}$ d) $\frac{9}{3}$

3. <u>True or false?</u> <u>Explain</u>. <u>Circle</u> the correct answer.

The smaller the denominator the greater the fraction.
You need to compare the numerators and the denominators.

b)   TRUE   (b) FALSE

Page 75

I need to add fractions:
$$\frac{1}{4} + \frac{5}{6} = ?$$

That's impossible!

One out of 4 equal parts   Five out of 6 equal parts

Just imagine, I ate $\frac{1}{4}$ of a Strawberry Chocolate bar and $\frac{3}{8}$ of a Hazelnut Chocolate bar. How much did I eat?

One out of 4 equal parts   Three out of 8 equal parts

Step 1: Find out equivalent fractions to get the same common denominator (the common denominator is LCD)

Step 2: Add the numerators only

Step 3: Simplify where possible

Yeah, first, we need to find out equivalent fractions!

$$\frac{1}{4} = \frac{2}{8} = \frac{3}{12} = \frac{4}{16} = \frac{5}{20} = \frac{6}{24} = \frac{7}{28} = \frac{8}{32}$$

$$\frac{3}{8} = \frac{6}{16} = \frac{9}{24} = \frac{12}{32}$$

Equivalent fractions are found out by multiplying both the numerator and denominator by the same number

Page 76

I have several like fractions with 8, 16, 24, 32 in the denominator. I colored the fractions strips grey.

$$\frac{1}{4} = \frac{2}{8} = \frac{3}{12} = \frac{4}{16} = \frac{5}{20} = \frac{6}{24} = \frac{7}{28} = \frac{8}{32}$$

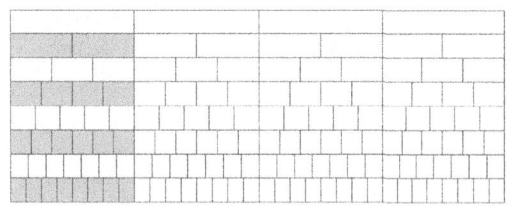

1 strip out of 4 strips is the same as 2 strips out of 8 strips, or 3 strips out of 12 strips, or 4 strips out of 16 strips, or 5 strips out of 20 strips, or 6 strips out of 24 strips, or 7 strips out of 28 strips, or 8 strips out of 32 strips - their value remains the same

$$\frac{3}{8} = \frac{6}{16} = \frac{9}{24} = \frac{12}{32}$$

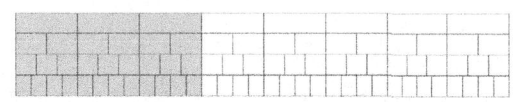

3 strips out of 8 strips are the same as 6 strips out of 16 strips, or 9 strips out of 24 strips, or 12 strips out of 32 strips - their value remains the same

Page 77

Aha! The denominators are the same in these fractions! These are common denominators.

Hint: The denominator (8) is the numeral that tells the number of parts a whole is divided into

$$\frac{2}{8} = \frac{4}{16} = \frac{6}{24} = \frac{8}{32}$$
$$\frac{3}{8} = \frac{6}{16} = \frac{9}{24} = \frac{12}{32}$$

Adding fractions is possible ONLY with first writing the fractions with common denominators

Step 1 is done! We changed the given fractions to equivalent fractions to find common denominators. Now we must find the least common denominator.

8 is the LEAST common denominator for $\frac{1}{4}$ ($=\frac{2}{8}$) and $\frac{3}{8}$.

These are like fractions since they both have 8 in the denominator. So, you can write 8 as a common denominator for both fractions. And then, add the numerators 2 and 3 to find the sum.

Step 1: Find the Common Denominator
Step 2: Add the Numerators to find the sum

$$\frac{1}{4} + \frac{3}{8} = \frac{2}{8} + \frac{3}{8} = \frac{2+3}{8} = \frac{5}{8}$$

Page 78

My problem is: $\frac{1}{4} + \frac{5}{6} = ?$

Don't worry, we'll start slow, I may have some ideas. Find the multiples!

Hint: To find the multiples of a number, multiply this number by the counting numbers (1, 2, 3, 4, etc.)

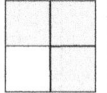

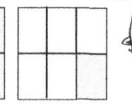

Step 1: Find the LEAST common multiple of 4 and 6.

Multiples of 4
For the denominator 4: $1 \times 4 = 4$ $2 \times 4 = 8$ $3 \times 4 = 12$
For the denominator 6: $1 \times 6 = 4$ $2 \times 6 = 12$ $3 \times 6 = 18$

Multiples of 6

The LEAST common multiple of 4 and 6 is 12:
$1 \times 4 = 4$ $2 \times 4 = 8$ $3 \times 4 = \boxed{12}$
$1 \times 6 = 4$ $2 \times 6 = \boxed{12}$ $3 \times 6 = 18$

How to find the least common multiple

Step 1: List all the multiples of the first denominator
Step 2: List all the multiples of the second denominator
Step 3: Circle the least multiple which is common to both denominators

Step 2: Draw a fractions bar, write 12 for the common denominator, and draw a slash above the numerator of each fraction:

$$\frac{1}{4} + \frac{5}{6} = \frac{\_ + \_}{12} =$$

Page 79

Step 3: Divide the least common multiple by each denominator and write the quotient above each slash:

Hint: Use slashes to make calculations easy and fun

$$\frac{1}{4}^{\backslash 3} + \frac{5}{6}^{\backslash 2} = \frac{\_ + \_}{12} =$$

$12 \div 4 = 3$
$12 \div 6 = 2$

Step 4: Multiply each numerator by the number above a slash:

$3 \times 1 = 3$
$2 \times 5 = 10$

$$\frac{1}{4}^{\backslash 3} + \frac{5}{6}^{\backslash 2} = \frac{(3 \times 1) + (2 \times 5)}{12} =$$

Hint: To add fractions, make sure that you add like fractions. That means they must have a common denominator

You got the "new" numerators. These are the numerators for the denominator 12:

$$\frac{1}{4} = \frac{3}{12}, \frac{5}{6} = \frac{10}{12}.$$

Step 5: Add the "new" numerators:

$$\frac{1}{4}^{\backslash 3} + \frac{5}{6}^{\backslash 2} = \frac{3 + 10}{12} = \frac{13}{12}$$

Page 80

1. <u>Add</u> fractions. <u>Write</u> the missing numbers. <u>Color</u> the fraction of each shape. <u>Simplify</u> if possible.

$\frac{1}{4} + \frac{1}{3} = ?$ $\hspace{2cm}$ $\frac{3}{4} + \frac{2}{5} = ?$

Step 1: Find the least common denominator

$\frac{1}{4} = \frac{2}{8} = \frac{3}{12}$ $\hspace{2cm}$ $\frac{3}{4} = \frac{6}{8} = \frac{9}{12} = \frac{12}{16} = \frac{15}{20}$

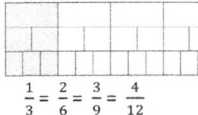

$\frac{1}{3} = \frac{2}{6} = \frac{3}{9} = \frac{4}{12}$

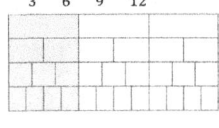

$\frac{2}{5} = \frac{4}{10} = \frac{6}{15} = \frac{8}{20}$

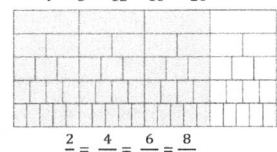

12 is the least common denominator $\hspace{1cm}$ 20 is the least common denominator

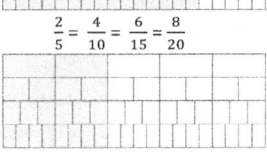

Step 2: Add the Numerators to find the sum

$\frac{1}{4} + \frac{1}{3} = \frac{3+4}{12} = \frac{7}{12} = \_\_$ $\hspace{1cm}$ $\frac{3}{4} + \frac{2}{5} = \frac{15+8}{20} = \frac{23}{20} = 1\frac{3}{20}$

2. <u>True or false?</u> <u>Explain</u>. <u>Circle</u> the correct answer.

The greater the numerator in like fractions the smaller the fraction.

The greater the numerator in like fractions the greater the fraction.

a) TRUE $\hspace{3cm}$ b) FALSE

Page 81

1. <u>Add</u> fractions. <u>Write</u> the missing numbers. <u>Color</u> the fraction of each shape. <u>Simplify</u> if possible.

$\frac{2}{3} + \frac{5}{12} = ?$ $\hspace{2cm}$ $\frac{5}{6} + \frac{3}{4} = ?$

Step 1: Find the least common denominator

$\frac{2}{3} = \frac{4}{6} = \frac{6}{9} = \frac{8}{12}$ $\hspace{2cm}$ $\frac{5}{6} = \frac{10}{12}$

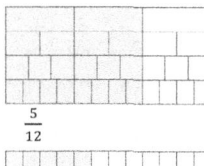

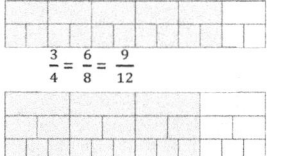

$\frac{3}{4} = \frac{6}{8} = \frac{9}{12}$

$\frac{5}{12}$

12 is the least common denominator $\hspace{1cm}$ 12 is the least common denominator

Step 2: Add the Numerators to find the sum

$\frac{2}{3} + \frac{5}{12} = \frac{8+5}{12} = \frac{13}{12} = 1\frac{1}{12}$ $\hspace{1cm}$ $\frac{5}{6} + \frac{3}{4} = \frac{10+9}{12} = \frac{19}{12} = 1\frac{7}{12}$

2. <u>True or false?</u> <u>Explain</u>. <u>Circle</u> the correct answer.

The greater the denominator in unlike fractions the greater the fraction.

You can compare only fractions with the like denominators.

a) TRUE $\hspace{3cm}$ b) FALSE

Page 82

1. <u>Add</u> fractions. <u>Write</u> the missing numbers. <u>Simplify</u> if possible.

$\frac{2}{6} + \frac{1}{2} = \frac{4+6}{12} = \frac{10}{12} = \frac{5}{6}$ $\hspace{1cm}$ $\frac{2}{4} + \frac{4}{5} = \frac{10+16}{20} = \frac{26}{20} = 1\frac{3}{10}$

$\frac{1}{3} + \frac{1}{4} = \frac{4+3}{12} = \frac{7}{12}$ $\hspace{1cm}$ $\frac{2}{3} + \frac{1}{6} = \frac{4+1}{6} = \frac{5}{6}$

$\frac{4}{6} + \frac{2}{4} = \frac{8+6}{12} = \frac{14}{12} = 1\frac{1}{6}$ $\hspace{1cm}$ $\frac{4}{9} + \frac{6}{18} = \frac{8+6}{18} = \frac{14}{18} = \frac{7}{9}$

$\frac{5}{14} + \frac{3}{7} = \frac{5+6}{14} = \frac{11}{14}$ $\hspace{1cm}$ $\frac{1}{6} + \frac{3}{8} = \frac{4+9}{24} = \frac{13}{24}$

$\frac{3}{4} + \frac{7}{16} = \frac{12+7}{16} = \frac{19}{16} = 1\frac{3}{16}$ $\hspace{1cm}$ $\frac{4}{5} + \frac{5}{6} = \frac{24+25}{30} = \frac{49}{30}$

$\frac{7}{12} + \frac{7}{8} = \frac{14+21}{24} = \frac{35}{24} = 1\frac{11}{24}$ $\hspace{1cm}$ $\frac{3}{10} + \frac{4}{6} = \frac{18+40}{60} = \frac{58}{60} = \frac{29}{30}$

Hint: Find the least common denominator by multiplying the greatest given denominator by 2, then, by 3, and so on until you get a number that is exactly divided by the given denominators.

2. <u>True or false?</u> <u>Explain</u>. <u>Circle</u> the correct answer.

The greater the denominator in unlike fractions the greater the fraction.

Any fractions can be compared by getting a common denominator.

a) TRUE $\hspace{3cm}$ b) FALSE

Page 83

1. <u>Add</u> fractions. <u>Write</u> the missing numbers. <u>Simplify</u> if possible.

$\frac{9}{16} + \frac{11}{12} = \frac{27+44}{48} = \frac{71}{48} = 1\frac{23}{48}$ $\hspace{1cm}$ $\frac{9}{18} + \frac{1}{4} = \frac{18+9}{36} = \frac{27}{36} = \frac{3}{4}$

$\frac{7}{20} + \frac{5}{8} = \frac{14+25}{40} = \frac{39}{40}$ $\hspace{1cm}$ $\frac{13}{18} + \frac{7}{10} = \frac{65+63}{90} = \frac{128}{90} = 1\frac{19}{45}$

$\frac{8}{25} + \frac{4}{15} = \frac{24+20}{75} = \frac{44}{75}$ $\hspace{1cm}$ $\frac{10}{28} + \frac{3}{4} = \frac{10+21}{28} = \frac{31}{28} = 1\frac{3}{28}$

$\frac{9}{20} + \frac{11}{14} + \frac{2}{7} = \frac{63+110+40}{140} = \frac{213}{140} = 1\frac{73}{140}$

$\frac{3}{7} + \frac{9}{14} + \frac{13}{21} = \frac{18+27+26}{42} = \frac{71}{42} = 1\frac{29}{42}$

$\frac{1}{2} + \frac{7}{13} + \frac{5}{26} = \frac{13+14+5}{26} = \frac{32}{26} = 1\frac{6}{13}$

2. <u>True or false?</u> <u>Explain</u>. <u>Circle</u> the correct answer.

The least common denominator is the least factor for the numerator.

The least common denominator is the least multiple of two numbers.

a) TRUE $\hspace{3cm}$ b) FALSE

Page 84

1. <u>Add</u> fractions. <u>Write</u> the missing numbers. <u>Simplify</u> if possible.

$\frac{2}{13} + \frac{5}{39} = \frac{6+5}{39} = \frac{11}{39}$ $\frac{2}{5} + \frac{3}{10} = \frac{4+3}{10} = \frac{7}{10}$

$\frac{5}{8} + \frac{5}{6} = \frac{15+20}{24} = \frac{35}{24} = 1\frac{11}{24}$ $\frac{5}{6} + \frac{7}{15} = \frac{25+14}{30} = \frac{39}{30} = 1\frac{3}{10}$

$\frac{7}{18} + \frac{8}{15} = \frac{35+48}{90} = \frac{83}{90}$ $\frac{5}{18} + \frac{3}{12} = \frac{30+27}{108} = \frac{57}{108} = \frac{19}{36}$

A common denominator may be found by multiplying together given denominators!

$\frac{11}{15} + \frac{7}{9} = 1\frac{23}{45}$ $\frac{9}{13} + \frac{1}{4} = \frac{49}{52}$

$\frac{13}{16} + \frac{7}{8} = 1\frac{11}{16}$ $\frac{3}{21} + \frac{1}{6} = \frac{13}{42}$

$\frac{5}{12} + \frac{3}{8} = \frac{19}{24}$ $\frac{7}{16} + \frac{5}{7} = 1\frac{17}{112}$

2. <u>True or false?</u> <u>Explain</u>. <u>Circle</u> the correct answer.

The mixed number is the sum of a proper fraction and an improper fraction.
_____

a) TRUE (circled)   b) FALSE

Page 85

I need to subtract fractions: $\frac{3}{4} - \frac{1}{6} = ?$

Let's try...

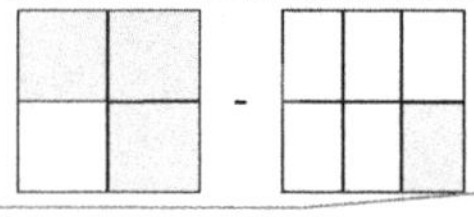

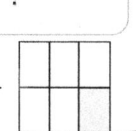

I ate $\frac{1}{6}$ out of $\frac{3}{4}$ of a Hazelnut Chocolate bar. How much is left? Can we apply the same steps to subtraction as we did in addition?

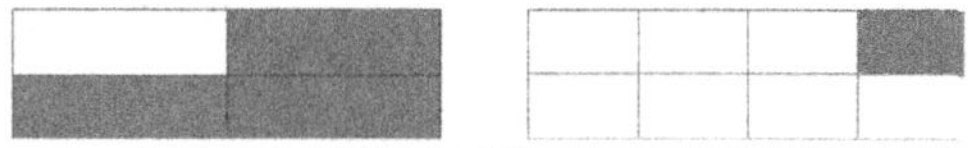

Step 1: Find out equivalent fractions to get the same common denominator (the common denominator is LCD)

Step 2: Subtract the numerators only

Step 3: Simplify where possible

Find out equivalent fractions!

$\frac{3}{4} = \frac{6}{8} = \frac{9}{12} = \frac{12}{16} = \frac{15}{20} = \frac{18}{24} = \frac{21}{28} = \frac{24}{32} = \frac{27}{36}$

$\frac{1}{6} = \frac{2}{12} = \frac{3}{18} = \frac{4}{24} = \frac{5}{30} = \frac{6}{36}$

Equivalent fractions are found out by multiplying both the numerator and denominator by the same number

Page 86

I have several like fractions with 12, 24, 36 in the denominator. I colored the fractions strips grey.

$\frac{3}{4} = \frac{6}{8} = \frac{9}{12} = \frac{12}{16} = \frac{15}{20} = \frac{18}{24} = \frac{21}{28} = \frac{24}{32} = \frac{27}{36}$

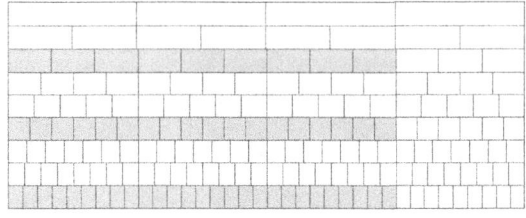

3 strips out of 4 strips are the same as 6 strips out of 8 strips, or 9 strips out of 12 strips, or 12 strips out of 16 strips, or 15 strips out of 20 strips, or 18 strips out of 24 strips, or 21 strips out of 28 strips, or 24 strips out of 32 strips, or 27 strips out of 36 strips - their value remains the same

$\frac{1}{6} = \frac{2}{12} = \frac{3}{18} = \frac{4}{24} = \frac{5}{30} = \frac{6}{36}$

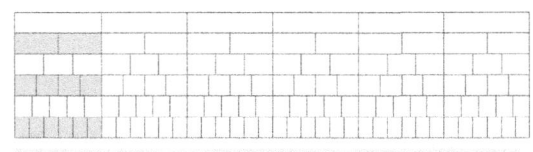

1 strip out of 6 strips is the same as 2 strips out of 12 strips, or 3 strips out of 18 strips, or 4 strips out of 24 strips, or 5 strips out of 30 strips, or 6 strips out of 36 strips - their value remains the same

Page 87

Aha! The denominators are the same in these fractions! These are common denominators

$\frac{9}{12} =$ $\frac{18}{24} =$ $\frac{27}{36}$
$\frac{2}{12} =$ $\frac{4}{24} =$ $\frac{6}{36}$

Hint: You have a list of equivalent fractions ☺

Step 1 is done! We changed the given fractions to equivalent fractions to find common denominators. Now we must find the least common denominator.

12 is the LEAST common denominator for $\frac{3}{4}$ $(= \frac{9}{12})$ and $\frac{1}{6}$ $(= \frac{2}{12})$.

Hint: The least common denominator is the denominator which is common to both denominators of the given fractions.

These are like fractions since they both have 12 in the denominator. So, you can write 12 as a common denominator for both fractions.

And then, subtract the numerators 9 and 2.

$\frac{3}{4} - \frac{1}{6} = \frac{9}{12} - \frac{2}{12} = \frac{9-2}{12} = \frac{7}{12}$

Page 88

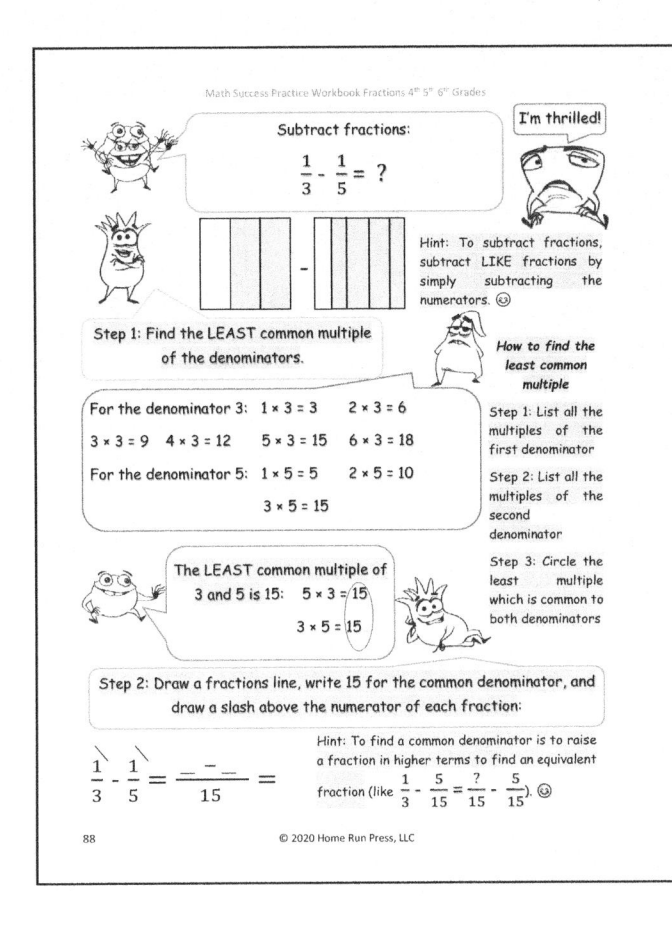

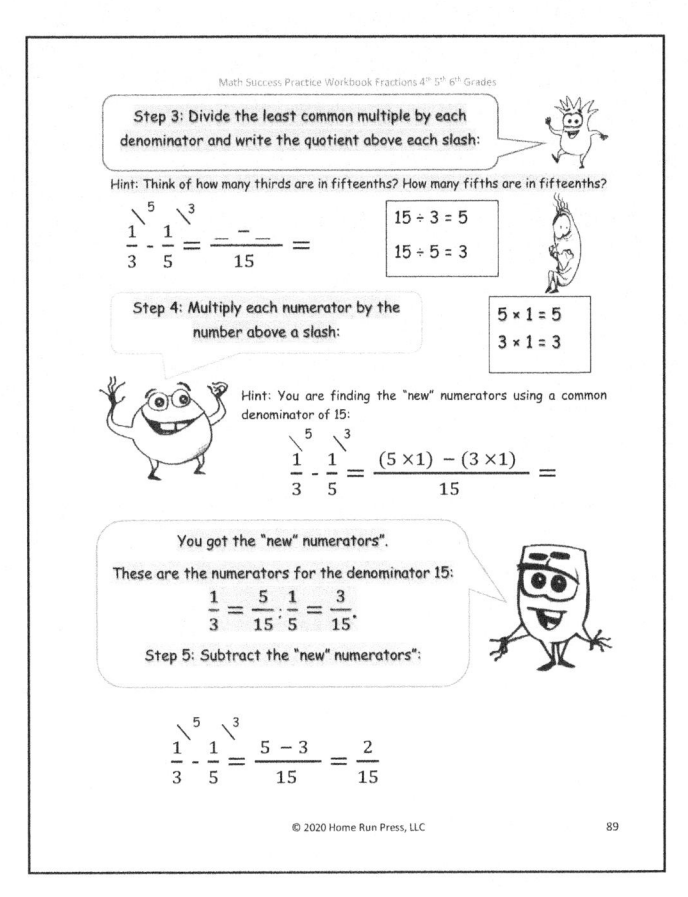

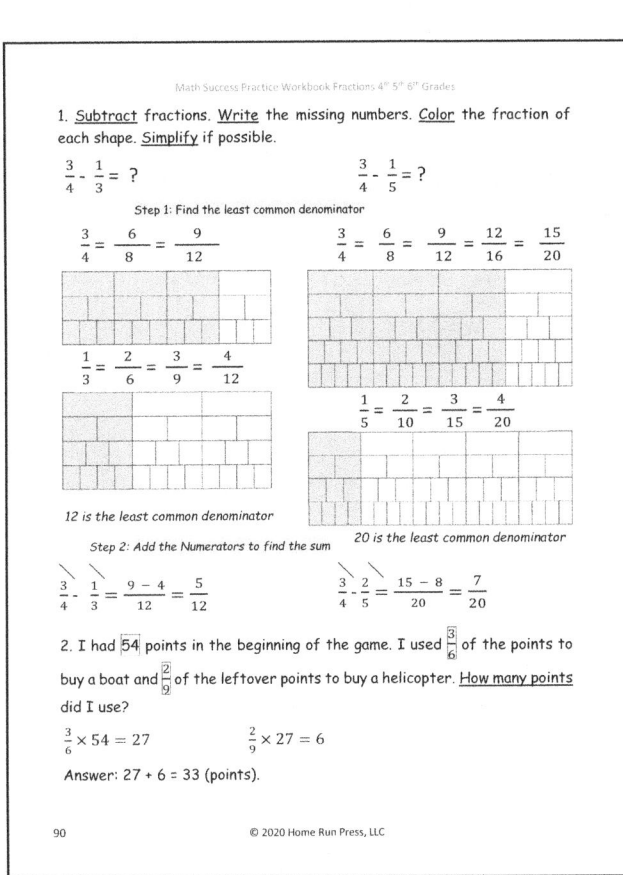

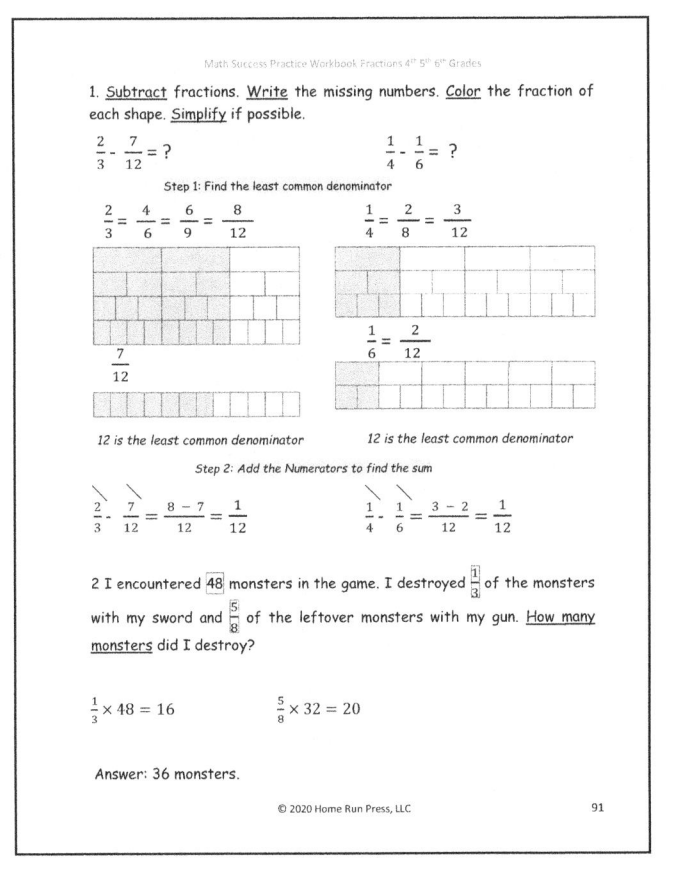

Subtract fractions:

$$\frac{1}{3} - \frac{1}{5} = ?$$

I'm thrilled!

Hint: To subtract fractions, subtract LIKE fractions by simply subtracting the numerators.

Step 1: Find the LEAST common multiple of the denominators.

For the denominator 3: $1 \times 3 = 3$ $2 \times 3 = 6$
$3 \times 3 = 9$ $4 \times 3 = 12$ $5 \times 3 = 15$ $6 \times 3 = 18$

For the denominator 5: $1 \times 5 = 5$ $2 \times 5 = 10$
$3 \times 5 = 15$

How to find the least common multiple
Step 1: List all the multiples of the first denominator
Step 2: List all the multiples of the second denominator
Step 3: Circle the least multiple which is common to both denominators

The LEAST common multiple of 3 and 5 is 15: $5 \times 3 = \boxed{15}$
$3 \times 5 = \boxed{15}$

Step 2: Draw a fractions line, write 15 for the common denominator, and draw a slash above the numerator of each fraction:

$$\frac{1}{3} - \frac{1}{5} = \frac{ - }{15} =$$

Hint: To find a common denominator is to raise a fraction in higher terms to find an equivalent fraction (like $\frac{1}{3} - \frac{5}{15} = \frac{?}{15} - \frac{5}{15}$).

Page 89

Step 3: Divide the least common multiple by each denominator and write the quotient above each slash:

Hint: Think of how many thirds are in fifteenths? How many fifths are in fifteenths?

$$\frac{1}{3}^{\diagdown 5} - \frac{1}{5}^{\diagdown 3} = \frac{-}{15} =$$

$15 \div 3 = 5$
$15 \div 5 = 3$

Step 4: Multiply each numerator by the number above a slash:

$5 \times 1 = 5$
$3 \times 1 = 3$

Hint: You are finding the "new" numerators using a common denominator of 15:

$$\frac{1}{3}^{\diagdown 5} - \frac{1}{5}^{\diagdown 3} = \frac{(5 \times 1) - (3 \times 1)}{15} =$$

You got the "new" numerators". These are the numerators for the denominator 15:

$$\frac{1}{3} = \frac{5}{15}, \frac{1}{5} = \frac{3}{15}.$$

Step 5: Subtract the "new" numerators":

$$\frac{1}{3}^{\diagdown 5} - \frac{1}{5}^{\diagdown 3} = \frac{5 - 3}{15} = \frac{2}{15}$$

Page 90

1. <u>Subtract</u> fractions. <u>Write</u> the missing numbers. <u>Color</u> the fraction of each shape. <u>Simplify</u> if possible.

$$\frac{3}{4} - \frac{1}{3} = ? \qquad \frac{3}{4} - \frac{1}{5} = ?$$

Step 1: Find the least common denominator

$$\frac{3}{4} = \frac{6}{8} = \frac{9}{12} \qquad \frac{3}{4} = \frac{6}{8} = \frac{9}{12} = \frac{12}{16} = \frac{15}{20}$$

$$\frac{1}{3} = \frac{2}{6} = \frac{3}{9} = \frac{4}{12} \qquad \frac{1}{5} = \frac{2}{10} = \frac{3}{15} = \frac{4}{20}$$

12 is the least common denominator   *20 is the least common denominator*

Step 2: Add the Numerators to find the sum

$$\frac{3}{4} - \frac{1}{3} = \frac{9-4}{12} = \frac{5}{12} \qquad \frac{3}{4} - \frac{2}{5} = \frac{15-8}{20} = \frac{7}{20}$$

2. I had $\boxed{54}$ points in the beginning of the game. I used $\frac{3}{6}$ of the points to buy a boat and $\frac{2}{9}$ of the leftover points to buy a helicopter. <u>How many points did I use?</u>

$\frac{3}{6} \times 54 = 27$ $\frac{2}{9} \times 27 = 6$

Answer: 27 + 6 = 33 (points).

Page 91

1. <u>Subtract</u> fractions. <u>Write</u> the missing numbers. <u>Color</u> the fraction of each shape. <u>Simplify</u> if possible.

$$\frac{2}{3} - \frac{7}{12} = ? \qquad \frac{1}{4} - \frac{1}{6} = ?$$

Step 1: Find the least common denominator

$$\frac{2}{3} = \frac{4}{6} = \frac{6}{9} = \frac{8}{12} \qquad \frac{1}{4} = \frac{2}{8} = \frac{3}{12}$$

$$\frac{7}{12} \qquad\qquad \frac{1}{6} = \frac{2}{12}$$

12 is the least common denominator   *12 is the least common denominator*

Step 2: Add the Numerators to find the sum

$$\frac{2}{3} - \frac{7}{12} = \frac{8-7}{12} = \frac{1}{12} \qquad \frac{1}{4} - \frac{1}{6} = \frac{3-2}{12} = \frac{1}{12}$$

2. I encountered $\boxed{48}$ monsters in the game. I destroyed $\frac{1}{3}$ of the monsters with my sword and $\frac{5}{8}$ of the leftover monsters with my gun. <u>How many monsters did I destroy?</u>

$\frac{1}{3} \times 48 = 16$ $\frac{5}{8} \times 32 = 20$

Answer: 36 monsters.

Page 92

1. Subtract fractions. Write the missing numbers. Simplify if possible.

$\frac{3}{8} - \frac{1}{6} = \frac{9-4}{24} = \frac{5}{24}$ $\frac{7}{9} - \frac{1}{3} = \frac{7-3}{9} = \frac{4}{9}$

$\frac{5}{8} - \frac{1}{4} = \frac{5-2}{8} = \frac{3}{8}$ $\frac{11}{15} - \frac{2}{5} = \frac{11-6}{15} = \frac{5}{15} = \frac{1}{3}$

$\frac{3}{10} - \frac{1}{4} = \frac{6-5}{20} = \frac{1}{20}$ $\frac{11}{16} - \frac{3}{8} = \frac{11-6}{16} = \frac{5}{16}$

$\frac{8}{12} - \frac{2}{6} = \frac{8-4}{12} = \frac{4}{12} = \frac{1}{3}$ $\frac{7}{8} - \frac{3}{6} = \frac{21-12}{24} = \frac{9}{24} = \frac{3}{8}$

$\frac{13}{18} - \frac{2}{3} = \frac{13-12}{18} = \frac{1}{18}$ $\frac{7}{15} - \frac{1}{3} = \frac{7-5}{15} = \frac{2}{15}$

$\frac{5}{7} - \frac{9}{14} = \frac{10-9}{14} = \frac{1}{14}$ $\frac{9}{10} - \frac{5}{6} = \frac{27-25}{30} = \frac{2}{30} = \frac{1}{15}$

2. Mother used $\frac{1}{3}$ of the water bucket to water the roses and $\frac{7}{24}$ of the water bucket to water the daisies, and $\frac{5}{16}$ of the water bucket to water the tulips. How much water (—) was left?

$\frac{1}{3} + \frac{7}{24} + \frac{5}{16} = \frac{16+14+15}{48} = \frac{45}{48} = \frac{15}{16}$

Answer: $\frac{1}{16}$

Page 93

1. Write in the missing numbers (add or subtract fractions).

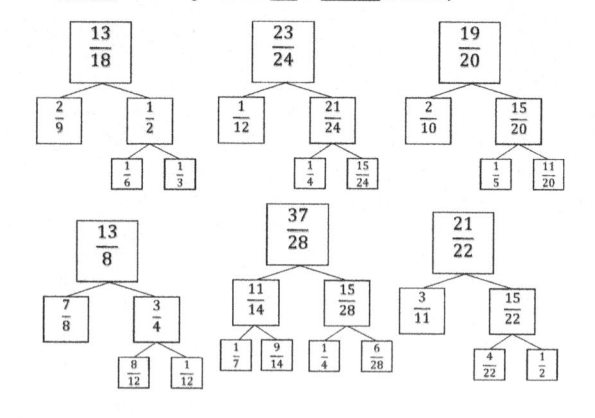

Tree 1: $\frac{13}{18}$ → $\frac{2}{9}$, $\frac{1}{2}$ → $\frac{1}{6}$, $\frac{1}{3}$

Tree 2: $\frac{23}{24}$ → $\frac{1}{12}$, $\frac{21}{24}$ → $\frac{1}{4}$, $\frac{15}{24}$

Tree 3: $\frac{19}{20}$ → $\frac{2}{10}$, $\frac{15}{20}$ → $\frac{1}{5}$, $\frac{11}{20}$

Tree 4: $\frac{13}{8}$ → $\frac{7}{8}$, $\frac{3}{4}$ → $\frac{8}{12}$, $\frac{1}{12}$

Tree 5: $\frac{37}{28}$ → $\frac{11}{14}$, $\frac{15}{28}$ → $\frac{1}{7}$, $\frac{9}{14}$, $\frac{1}{4}$, $\frac{6}{28}$

Tree 6: $\frac{21}{22}$ → $\frac{3}{11}$, $\frac{15}{22}$ → $\frac{4}{22}$, $\frac{1}{2}$

2. I spent $\frac{3}{7}$ of my allowance on a book and $\frac{5}{14}$ of it on a bag. I spent _____ of my allowance altogether. Circle the correct answer.

a) $\frac{10}{14}$ b) $\frac{9}{14}$ c) $\frac{14}{21}$ (d) $\frac{11}{14}$)

Page 94

1. Subtract fractions. Write the missing numbers. Simplify if possible.

$\frac{1}{2} - \frac{1}{6} = \frac{3-1}{6} = \frac{2}{6} = \frac{1}{3}$ $\frac{13}{30} - \frac{2}{5} = \frac{13-12}{30} = \frac{1}{30}$

$\frac{19}{24} - \frac{5}{8} = \frac{19-15}{24} = \frac{4}{24} = \frac{1}{6}$ $\frac{15}{28} - \frac{2}{4} = \frac{15-14}{28} = \frac{1}{28}$

$\frac{3}{4} - \frac{3}{12} = \frac{9-3}{12} = \frac{6}{12} = \frac{1}{2}$ $\frac{5}{7} - \frac{4}{9} = \frac{45-28}{63} = \frac{17}{63}$

$\frac{4}{11} - \frac{2}{22} = \frac{8-2}{22} = \frac{6}{22} = \frac{3}{11}$ $\frac{7}{10} - \frac{3}{40} = \frac{28-3}{40} = \frac{25}{40} = \frac{5}{8}$

$\frac{15}{20} - \frac{2}{5} = \frac{15-8}{20} = \frac{7}{20}$ $\frac{5}{6} - \frac{3}{8} = \frac{20-9}{24} = \frac{11}{24}$

$\frac{10}{15} - \frac{6}{12} = \frac{40-30}{60} = \frac{10}{60} = \frac{1}{6}$ $\frac{12}{36} - \frac{8}{48} = \frac{48-24}{144} = \frac{24}{144} = \frac{1}{6}$

2. Peter had some cookies. He ate $\frac{2}{9}$ of the cookies and he still had 21 cookies left. How many cookies did Peter have in the beginning?

$\frac{7}{9} \times x = 21$ → $\frac{1}{9} = 3$ → $\frac{9}{9} = 27$

Answer: 27 cookies.

Page 95

1. Subtract fractions. Write the missing numbers. Simplify if possible.

$\frac{5}{6} - \frac{3}{24} = \frac{17}{24}$ $\frac{7}{9} - \frac{1}{3} = \frac{4}{9}$

$\frac{5}{9} - \frac{1}{27} = \frac{14}{27}$ $\frac{7}{16} - \frac{5}{48} = \frac{1}{3}$

$\frac{23}{24} - \frac{1}{3} - \frac{1}{4} = \frac{23-8-6}{24} = \frac{3}{8}$

$\frac{5}{6} - \frac{1}{5} - \frac{1}{15} = \frac{25-6-2}{30} = \frac{17}{30}$

$\frac{6}{7} - \frac{2}{5} - \frac{5}{70} = \frac{60-28-5}{70} = \frac{27}{70}$

$\frac{11}{12} - \frac{5}{48} - \frac{7}{24} = \frac{44-5-14}{48} = \frac{25}{48}$

2. Kelly had some cards. She traded $\frac{3}{8}$ of the cards and she still had 35 cards left. How many cards did Kelly have in the beginning?

$\frac{5}{8} \times x = 35$ → $\frac{1}{8} = 7$ → $\frac{8}{8} = 56$

Answer: 56 cards.

Page 96

1. Write in the missing numbers (add or subtract).

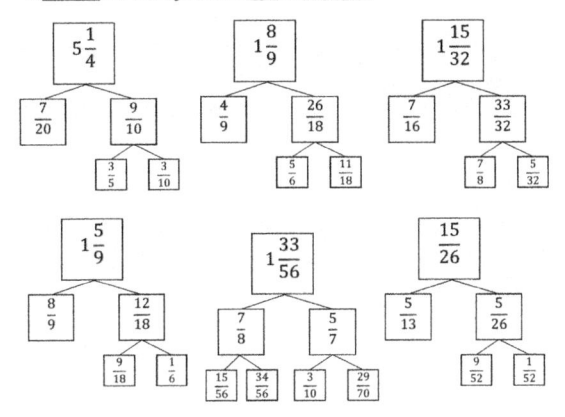

```
        [5 1/4]                    [1 8/9]                   [1 15/32]
       /      \                   /      \                  /        \
    [7/20]  [9/10]             [4/9]   [26/18]           [7/16]   [33/32]
            /    \                     /    \                     /     \
         [3/5]  [3/10]              [5/6]  [11/18]            [7/8]   [5/32]


        [1 5/9]                   [1 33/56]                 [15/26]
       /      \                   /       \                /       \
    [8/9]   [12/18]            [7/8]    [5/7]           [5/13]   [5/26]
            /    \           /    \     /    \                   /    \
         [9/18] [1/6]    [15/56][34/56][3/10][29/70]          [9/52] [1/52]
```

2. The length of a rectangle is $\frac{1}{2}$ inch shorter than its width. If its width is 3 inches, find its area. Circle the correct answer.

a) 7   b) $\frac{16}{2}$ c) $6\frac{1}{2}$ (d) $7\frac{1}{2}$)

Page 97

1. Subtract fractions. Write the missing numbers. Simplify if possible.

$1 - \frac{1}{6} = \frac{6}{6} - \frac{1}{6} = \frac{6-1}{6} = \frac{5}{6}$ $1 - \frac{1}{3} = \frac{3}{3} - \frac{1}{3} = \frac{3-1}{3} = \frac{2}{3}$

Hint: To subtract a fraction from a whole number: change a whole number to a fraction with the same denominator as that in the fraction to be subtracted

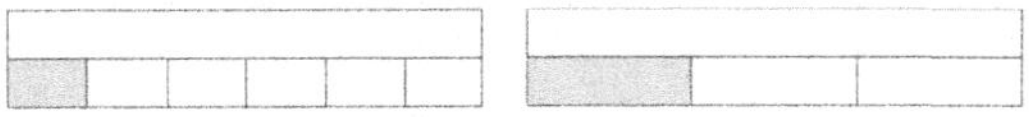

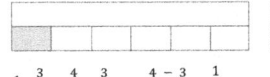

$1 - \frac{3}{4} = \frac{4}{4} - \frac{3}{4} = \frac{4-3}{4} = \frac{1}{4}$ $1 - \frac{2}{5} = \frac{5}{5} - \frac{2}{5} = \frac{5-2}{5} = \frac{3}{5}$

Hint: Rewrite a whole number to a fraction so that both fractions have a common denominator:

$1 - \frac{5}{7} = \frac{7}{7} - \frac{5}{7} = \frac{7-5}{7} = \frac{2}{7}$ $1 - \frac{3}{8} = \frac{8}{8} - \frac{3}{8} = \frac{8-3}{8} = \frac{5}{8}$

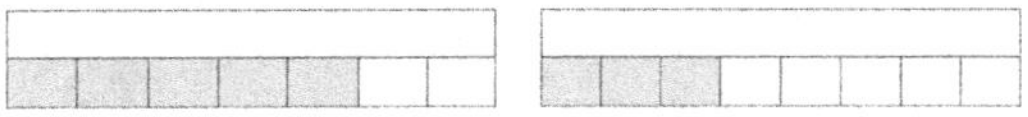

Hint: First, simplify if possible! Then, subtract!

$2 - \frac{2}{6} = \frac{12}{6} - \frac{2}{6} = \frac{12-2}{6} = \frac{10}{6}$ $3 - \frac{3}{6} = \frac{18}{6} - \frac{3}{6} = \frac{18-3}{6} = \frac{15}{6}$

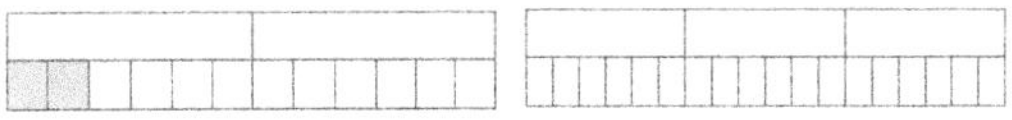

$4 - \frac{8}{12} = \frac{12}{3} - \frac{2}{3} = \frac{12-2}{3} = \frac{10}{3}$ $5 - \frac{9}{15} = 5 - \frac{3}{5} = \frac{25-3}{5} = \frac{22}{5}$

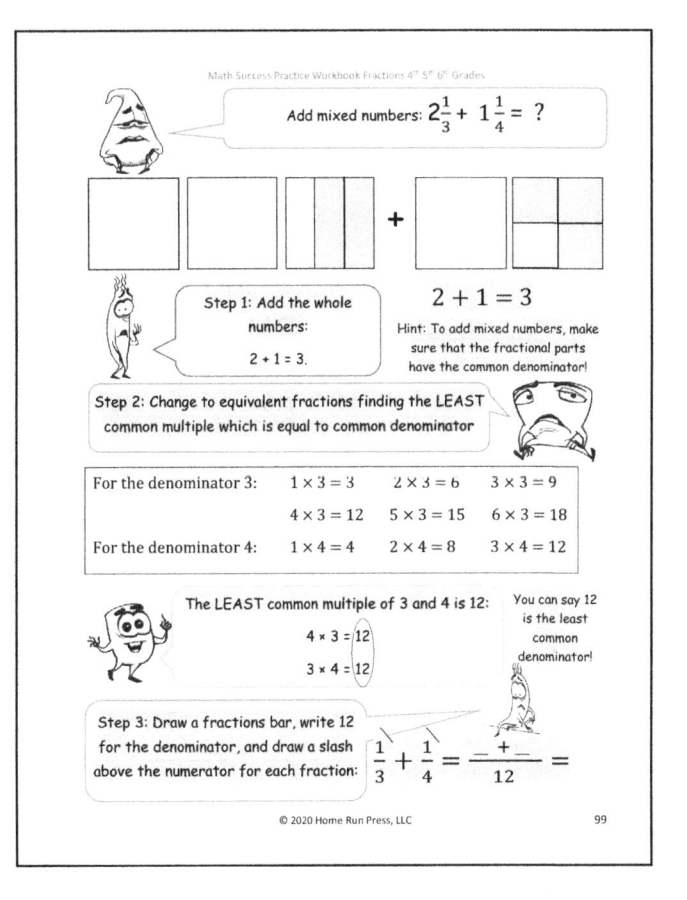

Page 98

1. Subtract fractions. Write the missing numbers. Simplify if possible.

Hint: To subtract a fraction from a whole number: rewrite a whole number to a mixed number: a whole number and a fraction with the same denominator (like $2\frac{9}{9}$).

$3 - \frac{4}{9} = 2\frac{9}{9} - \frac{4}{9} = 2\frac{9-4}{9} = 2\frac{5}{9}$ $6 - \frac{8}{12} = 5\frac{12}{12} - \frac{8}{12} = 5\frac{12-8}{12} = 5\frac{1}{3}$

$4 - \frac{2}{3} = 3\frac{3}{3} - \frac{2}{3} = 3\frac{3-2}{3} = 3\frac{1}{3}$ $5 - \frac{2}{5} = 4\frac{5}{5} - \frac{2}{5} = 4\frac{5-2}{5} = 4\frac{3}{5}$

$6 - \frac{9}{14} = 5\frac{14}{14} - \frac{9}{14} = 5\frac{14-9}{14} = 5\frac{5}{14}$ $7 - \frac{5}{6} = 6\frac{6}{6} - \frac{5}{6} = 6\frac{6-5}{6} = 6\frac{1}{6}$

2. Multiply and simplify if possible. Circle the correct answer (letter).

Find $\frac{10}{18}$ of 54.

$\frac{10}{18} \times 54 = \frac{10 \times 54}{18 \times 1} = 30$

a) 50
b) 44
(c) 30)

Find $\frac{2}{3}$ of 72.

Step 1: 72 ÷ 3 = 24
Step 2: 2 × 24 = 48

(a) 48)
b) 52
c) 60

Find $\frac{3}{4}$ of 56.

Step 1: 56 ÷ 4 = 14
Step 2: 3 × 14 = 42

d) 32
e) 40
(f) 42)

Page 99

Add mixed numbers: $2\frac{1}{3} + 1\frac{1}{4} = ?$

$2 + 1 = 3$

Hint: To add mixed numbers, make sure that the fractional parts have the common denominator!

Step 1: Add the whole numbers:
$2 + 1 = 3$.

Step 2: Change to equivalent fractions finding the LEAST common multiple which is equal to common denominator

For the denominator 3: $1 \times 3 = 3$ $2 \times 3 = 6$ $3 \times 3 = 9$
                         $4 \times 3 = 12$ $5 \times 3 = 15$ $6 \times 3 = 18$

For the denominator 4: $1 \times 4 = 4$ $2 \times 4 = 8$ $3 \times 4 = 12$

The LEAST common multiple of 3 and 4 is 12: $4 \times 3 = \boxed{12}$ $3 \times 4 = \boxed{12}$

You can say 12 is the least common denominator!

Step 3: Draw a fractions bar, write 12 for the denominator, and draw a slash above the numerator for each fraction: $\frac{1}{3} + \frac{1}{4} = \frac{\_ + \_}{12} =$

Page 100

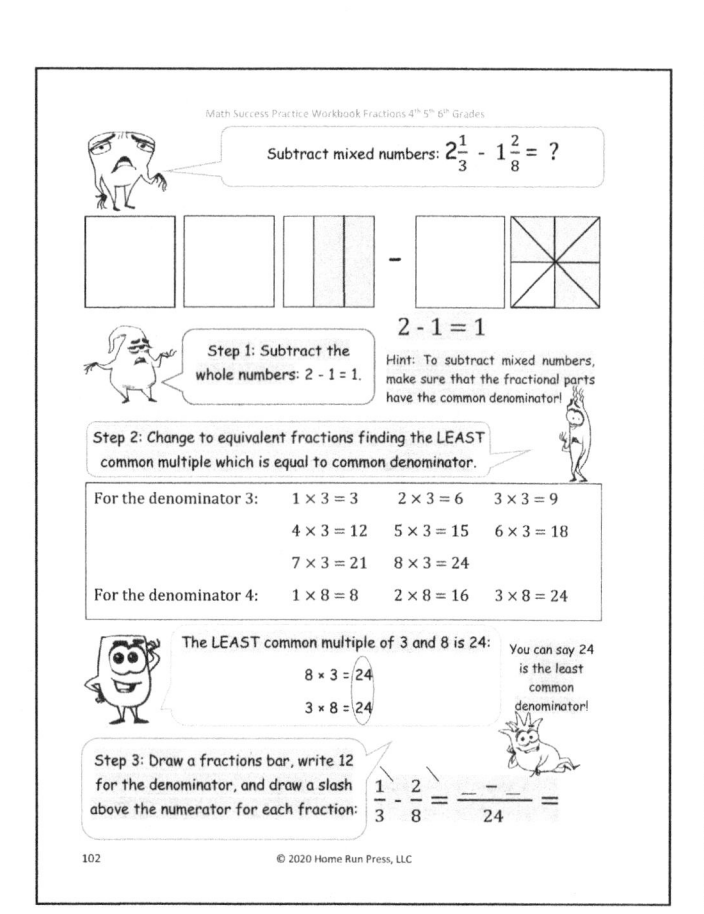

Step 4: Divide the least common multiple by each denominator and write the quotient above each slash:

Hint: Use slashes in the beginning to make the process easy and fun ☺

$$\frac{1}{3}\overset{4}{\diagup} + \frac{1}{4}\overset{3}{\diagup} = \frac{\_ + \_}{12} =$$

$12 \div 3 = 4$
$12 \div 4 = 3$

Step 5: Multiply each numerator by the number above a slash:

$4 \times 1 = 4$
$3 \times 1 = 3$

Hint: You can also say that you're finding equivalent fractions using a common denominator of 12 ☺

$$\frac{1}{3}\overset{4}{\diagup} + \frac{1}{4}\overset{3}{\diagup} = \frac{(4 \times 1) + (3 \times 1)}{12} =$$

You got the "new" numerators.
These are the numerators for the denominator of 12:

$$\frac{1}{3} = \frac{4}{12}, \frac{1}{4} = \frac{3}{12}.$$

Step 6: Add the "new" numerators:

$$\frac{1}{3}\overset{4}{\diagup} + \frac{1}{4}\overset{3}{\diagup} = \frac{4 + 3}{12} = \frac{7}{12}$$

Step 7: Add the whole number and a fractional part:

$$3 + \frac{7}{12} = 3\frac{7}{12}$$

Page 101

1. <u>Add</u> fractions. <u>Write</u> the missing numbers. <u>Simplify</u> if possible.

$$2\frac{3}{8} + 4\frac{5}{9} = 6\frac{67}{72}$$

1) <u>Add</u> whole numbers
2) <u>Add</u> fractions

1) 2 + 4 = 6
2) $\frac{3}{8} + \frac{5}{9} = \frac{27 + 40}{72} = \frac{67}{72}$

$$3\frac{3}{4} + 5\frac{3}{5} = 9\frac{7}{20}$$

1) 3 + 5 = 8
2) $\frac{3}{4} + \frac{3}{5} = \frac{15 + 12}{20} = \frac{27}{20} = 1\frac{7}{20}$

$$6\frac{3}{10} + 4\frac{5}{6} = 11\frac{2}{15}$$

1) 6 + 4 = 10
2) $\frac{3}{10} + \frac{5}{6} = \frac{9 + 25}{30} = \frac{34}{30} = 1\frac{2}{15}$

2. <u>Change</u> the mixed number to an improper faction. <u>Circle</u> the correct answer.

$$3\frac{7}{13} = \frac{46}{13} \qquad\qquad 8\frac{5}{9} = \frac{77}{9}$$

a) $\frac{39}{13}$ b) $\frac{23}{13}$ (c) $\frac{46}{13}$)   a) $\frac{72}{9}$ b) $\frac{75}{9}$ (c) $\frac{77}{9}$)

Page 102

Subtract mixed numbers: $2\frac{1}{3} - 1\frac{2}{8} = ?$

Step 1: Subtract the whole numbers: 2 - 1 = 1.

2 - 1 = 1

Hint: To subtract mixed numbers, make sure that the fractional parts have the common denominator!

Step 2: Change to equivalent fractions finding the LEAST common multiple which is equal to common denominator.

For the denominator 3:
$1 \times 3 = 3$ $2 \times 3 = 6$ $3 \times 3 = 9$
$4 \times 3 = 12$ $5 \times 3 = 15$ $6 \times 3 = 18$
$7 \times 3 = 21$ $8 \times 3 = 24$

For the denominator 4: $1 \times 8 = 8$ $2 \times 8 = 16$ $3 \times 8 = 24$

The LEAST common multiple of 3 and 8 is 24:
$8 \times 3 = \boxed{24}$
$3 \times 8 = \boxed{24}$

You can say 24 is the least common denominator!

Step 3: Draw a fractions bar, write 12 for the denominator, and draw a slash above the numerator for each fraction:

$$\frac{1}{3}\diagup - \frac{2}{8}\diagup = \frac{\_ - \_}{24} =$$

Page 103

Step 4: Divide the least common multiple by each denominator and write the quotient above each slash:

Hint: Use slashes in the beginning to make the process easy and fun ☺

$$\frac{1}{3}\overset{8}{\diagup} - \frac{2}{8}\overset{3}{\diagup} = \frac{\_ - \_}{24} =$$

$24 \div 3 = 8$
$24 \div 8 = 3$

Step 5: Multiply each numerator by the number above a slash:

$8 \times 1 = 8$
$3 \times 2 = 6$

Hint: You can also say that you're finding the equivalent fractions using a common denominator of 24 ☺

Your "new" numerators

$$\frac{1}{3}\overset{8}{\diagup} - \frac{2}{8}\overset{3}{\diagup} = \frac{(8 \times 1) - (3 \times 2)}{24} =$$

You got the "new" numerators.
These are the numerators for the denominator 12:

$$\frac{1}{3} = \frac{8}{24}, \frac{2}{8} = \frac{6}{24}.$$

Step 6: Add the "new" numerators" and simplify:

$$\frac{1}{3}\overset{8}{\diagup} - \frac{2}{8}\overset{3}{\diagup} = \frac{8 - 6}{24} = \frac{2}{24} = \frac{1}{12}$$

Step 7: Add the whole number and a fractional part:

$$1 + \frac{1}{12} = 1\frac{1}{12}$$

Page 104

1. <u>Subtract</u> fractions. <u>Write</u> the missing numbers. <u>Simplify</u> if possible.

$3\frac{5}{8} - 1\frac{5}{6} = 1\frac{19}{24}$

1) <u>Subtract</u> whole numbers   2) <u>Subtract</u> fractions

To subtract a greater fraction from a smaller fraction: Step 1: borrow one from a whole number; Step 2: rewrite the borrowed whole number to a fraction: $3\frac{1}{3} - 1\frac{3}{8} = 1\frac{23}{24}$

Step 1: $3 - 1 = \cancel{2}\,1$ Step 2: $\frac{1}{3} - \frac{3}{8} = \frac{8-9}{24} = \frac{8+24-9}{24} = \frac{23}{24}$

1) $3 - 1 = \cancel{2}\,1$ 2) $\frac{5}{8} - \frac{5}{6} = \frac{15+24-20}{24} = \frac{19}{24}$

$8\frac{7}{10} - 2\frac{3}{4} = 5\frac{19}{20}$

1) $8 - 2 = \cancel{6}\,5$ 2) $\frac{7}{10} - \frac{3}{4} = \frac{14+20-15}{20} = \frac{19}{20}$

$9\frac{1}{15} - 3\frac{7}{9} = 5\frac{13}{45}$

1) $9 - 3 = \cancel{6}\,5$ 2) $\frac{1}{15} - \frac{7}{9} = \frac{3+45-35}{45} = \frac{13}{45}$

2. <u>Change</u> the improper faction to a mixed number. <u>Circle</u> the correct answer.

$\frac{32}{5} = 6\frac{2}{5}$

a) $3\frac{2}{5}$ b) $6\frac{1}{5}$ c) $\boxed{6\frac{2}{5}}$

$\frac{49}{11} = 4\frac{5}{11}$

a) $11\frac{5}{11}$ b) $1\frac{38}{11}$ c) $\boxed{4\frac{5}{11}}$

Page 105

1. <u>Multiply</u> fractions. <u>Write</u> the missing numbers. <u>Simplify</u> if possible.

Step 1: <u>Change</u> the mixed number to an improper fraction
Step 2: <u>Multiply</u> the numerators and denominators

$3 \times 2\frac{3}{15} = \frac{3}{1} \times \frac{33}{15} = \frac{3}{1} \times \frac{11}{5} = \frac{11+11+11}{5} = 6\frac{3}{5}$

Hint: Simplify before multiplying where you can! (cancel 3s from 33 and 15)

Hint: Change the mixed number to improper fractions!

$6 \times 3\frac{5}{12} = \frac{6}{1} \times \frac{41}{12} = \frac{6 \times 41}{1 \times 12} = \frac{41}{2} = 20\frac{1}{2}$

Hint: Cancel or cross out the common factors of the numerator and the denominator to simplify where it's possible ☺

$5 \times 2\frac{7}{20} = \frac{5}{1} \times \frac{47}{20} = \frac{5 \times 47}{1 \times 20} = \frac{47}{4} = 11\frac{3}{4}$

$4 \times 5\frac{5}{16} = \frac{4}{1} \times \frac{85}{16} = \frac{4 \times 85}{1 \times 16} = \frac{85}{4} = 21\frac{1}{4}$

2. <u>Multiply</u>. <u>Express</u> your answer in simplest form.

$\frac{9}{16} \times \frac{20}{36}$

a) $\frac{15}{36}$ b) $\boxed{\frac{5}{16}}$ c) $\frac{180}{432}$ d) $2\frac{2}{5}$

Page 106

1. <u>Multiply</u> fractions. <u>Write</u> the missing numbers. <u>Simplify</u> if possible.

Step 1: <u>Change</u> the mixed number to an improper fraction
Step 2: <u>Multiply</u> the numerators and denominators

$4\frac{1}{5} \times 4\frac{2}{7} = \frac{21}{5} \times \frac{30}{7} = \frac{21 \times 30}{5 \times 7} = \frac{18}{1} = 18$

Hint: Factor the numerator and the denominator by the same number to reduce the fraction to lowest terms! ☺

$2\frac{1}{4} \times 3\frac{4}{6} = \frac{9}{4} \times \frac{22}{6} = \frac{9 \times 22}{4 \times 6} = \frac{33}{4} = 8\frac{1}{4}$

$3\frac{1}{2} \times 2\frac{2}{7} = \frac{7}{2} \times \frac{16}{7} = \frac{7 \times 16}{2 \times 7} = 8$

$9\frac{5}{8} \times 3\frac{7}{11} = \frac{77}{8} \times \frac{40}{11} = \frac{77 \times 40}{8 \times 11} = 35$

2. <u>Multiply</u>. <u>Express</u> your answer in simplest form.

$\frac{5}{12} \times \frac{36}{55}$

a) $\frac{3}{10}$ b) $\frac{36}{60}$ c) $\boxed{\frac{3}{11}}$ d) $1\frac{2}{5}$

Page 107

1. <u>Write in</u> the missing numbers (practice adding and subtracting mixed numbers).

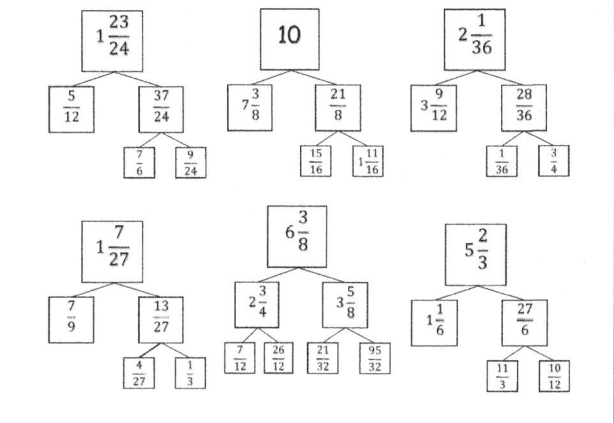

$1\frac{23}{24}$ → $\frac{5}{12}$, $\frac{37}{24}$ → $\frac{7}{6}$, $\frac{9}{24}$

10 → $7\frac{3}{8}$, $\frac{21}{8}$ → $\frac{15}{16}$, $1\frac{11}{16}$

$2\frac{1}{36}$ → $3\frac{9}{12}$, $\frac{28}{36}$ → $\frac{1}{36}$, $\frac{3}{4}$

$1\frac{7}{27}$ → $\frac{7}{9}$, $\frac{13}{27}$ → $\frac{4}{27}$, $\frac{1}{3}$

$6\frac{3}{8}$ → $2\frac{3}{4}$, $3\frac{5}{8}$ → $\frac{7}{12}$, $\frac{26}{12}$, $\frac{21}{32}$, $\frac{95}{32}$

$5\frac{2}{3}$ → $1\frac{1}{6}$, $\frac{27}{6}$ → $\frac{11}{3}$, $\frac{10}{12}$

2. A class is $\frac{5}{8}$ filled with spectators. $\frac{2}{5}$ of them are boys and the rest are girls. The fraction of girls in the class is _____.

a) $1\frac{1}{40}$ B) $\frac{3}{5}$ c) $\boxed{\frac{9}{40}}$ d) $\frac{3}{8}$

Page 108

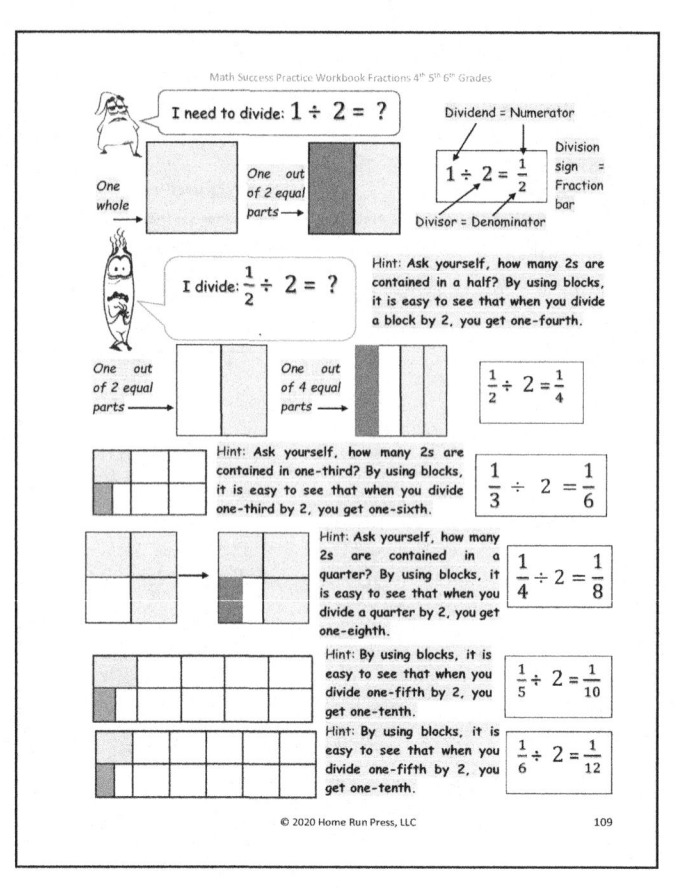

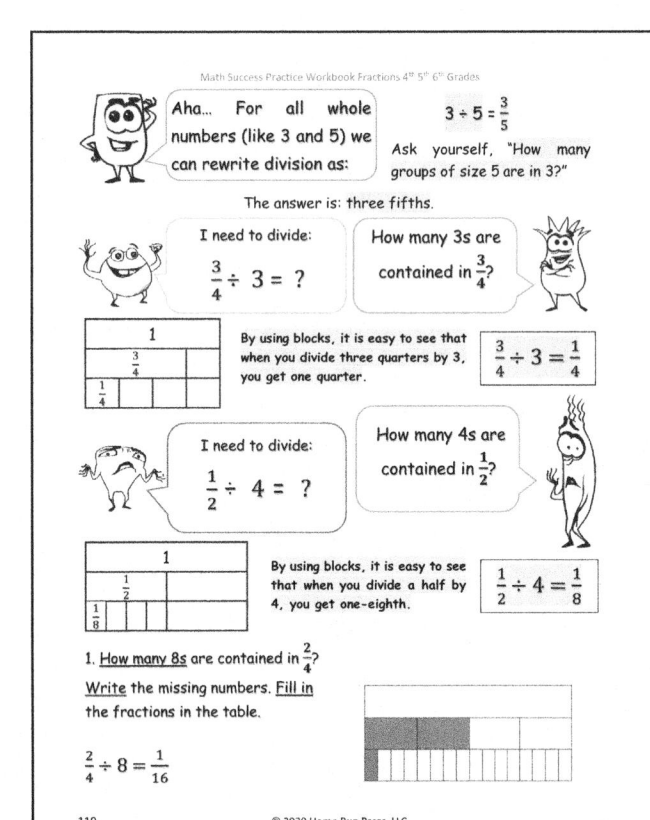

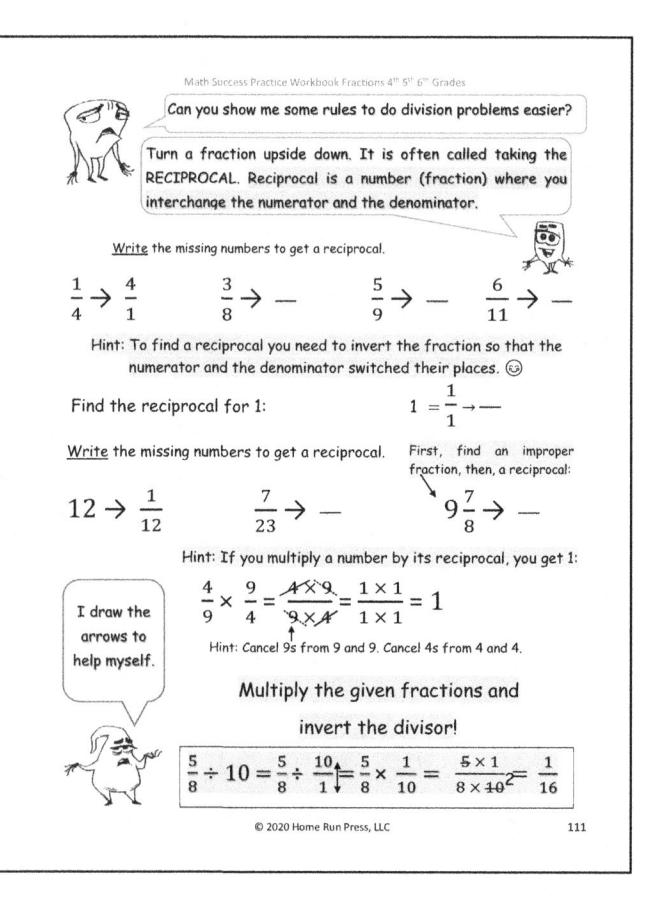

1. <u>Divide</u> fractions. <u>Write</u> the missing numbers. <u>Simplify</u> if possible.

Divide fractions! That's impossible! I cannot divide 3 blocks by, say, $\frac{5}{9}$!

$3 \div \frac{5}{9} = $ **no** answer, no solution, let's skip it.

Your problem means: "How many $\frac{5}{9}$'s are there in 3?" or "How many $\frac{5}{9}$'s are contained in 3?", or "How many $\frac{5}{9}$'s in 3?"

Step 1: Change 3 to $\frac{1}{9}$'s. Use a common denominator of 9. $\quad 3 = \frac{27}{9}$

Step 2: Divide the numerators, then, divide the denominators, then, divide the numerator by the denominator.

So, $3 \div \frac{5}{9} = \frac{27}{9} \div \frac{5}{9} = \frac{(27 \div 5)}{(9 \div 9)} = \frac{(27 \div 5)}{1} = \frac{27}{5} = 5\frac{2}{5}$

Denominator $5\overline{)27}$ =Numerator, $\;5$ =The whole number, 25, 2 = Remainder

! Dividend = Numerator
Divisor = Denominator
Quotient = The Whole Number
Remainder = Numerator of a fraction part

$5 \div \frac{2}{7} = \frac{35}{7} \div \frac{2}{7} = \frac{(35 \div 2)}{(7 \div 7)} = \frac{35}{2} = 17\frac{1}{2}$

$4 \div \frac{2}{3} = \frac{12}{3} \div \frac{2}{3} = \frac{(12 \div 2)}{(3 \div 3)} = \frac{12}{2} = 6$

Page 109

I need to divide: $1 \div 2 = ?$

Dividend = Numerator
$1 \div 2 = \frac{1}{2}$
Division sign = Fraction bar
Divisor = Denominator

One whole | One out of 2 equal parts

I divide: $\frac{1}{2} \div 2 = ?$

Hint: Ask yourself, how many 2s are contained in a half? By using blocks, it is easy to see that when you divide a block by 2, you get one-fourth.

One out of 2 equal parts → One out of 4 equal parts → $\frac{1}{2} \div 2 = \frac{1}{4}$

Hint: Ask yourself, how many 2s are contained in one-third? By using blocks, it is easy to see that when you divide one-third by 2, you get one-sixth. $\quad \frac{1}{3} \div 2 = \frac{1}{6}$

Hint: Ask yourself, how many 2s are contained in a quarter? By using blocks, it is easy to see that when you divide a quarter by 2, you get one-eighth. $\quad \frac{1}{4} \div 2 = \frac{1}{8}$

Hint: By using blocks, it is easy to see that when you divide one-fifth by 2, you get one-tenth. $\quad \frac{1}{5} \div 2 = \frac{1}{10}$

Hint: By using blocks, it is easy to see that when you divide one-fifth by 2, you get one-tenth. $\quad \frac{1}{6} \div 2 = \frac{1}{12}$

Page 110

Aha... For all whole numbers (like 3 and 5) we can rewrite division as:

$3 \div 5 = \frac{3}{5}$

Ask yourself, "How many groups of size 5 are in 3?"

The answer is: three fifths.

I need to divide: $\frac{3}{4} \div 3 = ?$

How many 3s are contained in $\frac{3}{4}$?

By using blocks, it is easy to see that when you divide three quarters by 3, you get one quarter. $\quad \frac{3}{4} \div 3 = \frac{1}{4}$

I need to divide: $\frac{1}{2} \div 4 = ?$

How many 4s are contained in $\frac{1}{2}$?

By using blocks, it is easy to see that when you divide a half by 4, you get one-eighth. $\quad \frac{1}{2} \div 4 = \frac{1}{8}$

1. <u>How many 8s</u> are contained in $\frac{2}{4}$? <u>Write</u> the missing numbers. <u>Fill in</u> the fractions in the table.

$\frac{2}{4} \div 8 = \frac{1}{16}$

Page 111

Can you show me some rules to do division problems easier?

Turn a fraction upside down. It is often called taking the RECIPROCAL. Reciprocal is a number (fraction) where you interchange the numerator and the denominator.

<u>Write</u> the missing numbers to get a reciprocal.

$\frac{1}{4} \rightarrow \frac{4}{1} \qquad \frac{3}{8} \rightarrow \underline{} \qquad \frac{5}{9} \rightarrow \underline{} \qquad \frac{6}{11} \rightarrow \underline{}$

Hint: To find a reciprocal you need to invert the fraction so that the numerator and the denominator switched their places. ☺

Find the reciprocal for 1: $\qquad 1 = \frac{1}{1} \rightarrow \underline{}$

<u>Write</u> the missing numbers to get a reciprocal. First, find an improper fraction, then, a reciprocal:

$12 \rightarrow \frac{1}{12} \qquad \frac{7}{23} \rightarrow \underline{} \qquad 9\frac{7}{8} \rightarrow \underline{}$

Hint: If you multiply a number by its reciprocal, you get 1:

$\frac{4}{9} \times \frac{9}{4} = \frac{\cancel{4} \times \cancel{9}}{\cancel{9} \times \cancel{4}} = \frac{1 \times 1}{1 \times 1} = 1$

I draw the arrows to help myself.

Hint: Cancel 9s from 9 and 9. Cancel 4s from 4 and 4.

Multiply the given fractions and invert the divisor!

$\frac{5}{8} \div 10 = \frac{5}{8} \div \frac{10}{1} = \frac{5}{8} \times \frac{1}{10} = \frac{5 \times 1}{8 \times \cancel{10}^2} = \frac{1}{16}$

Page 112

1. Divide fractions. Write the missing numbers. Simplify if possible.

$\frac{10}{3} \div 2 = \frac{10}{3} \div \frac{2}{1} = \frac{10}{3} \times \frac{1}{2} = \frac{10 \times 1}{3 \times 2} = 1\frac{2}{3}$

$\frac{8}{15} \div 4 = \frac{8}{15} \div \frac{4}{1} = \frac{8}{15} \times \frac{1}{4} = \frac{8 \times 1}{15 \times 4} = \frac{2}{15}$

$\frac{2}{3} \div 4 = \frac{2}{3} \div \frac{4}{1} = \frac{2}{3} \times \frac{1}{4} = \frac{2 \times 1}{3 \times 4} = \frac{1}{6}$

$\frac{5}{6} \div 10 = \frac{1}{12}$ $\frac{13}{18} \div 26 = \frac{1}{36}$

2. The perimeter of a rectangle is $1\frac{5}{12}$ in². Find the length of the rectangle if its width is $\frac{3}{8}$ in long.

P = 2L + 2W

$1\frac{5}{12} = 2 \times L + 2 \times \frac{3}{8}$

$L = (1\frac{5}{12} - \frac{3}{8}) \div 2$

$L = \frac{1}{3}$

Answer: $\frac{3}{8}$ in long.

Page 113

1. Divide fractions. Write the missing numbers. Simplify if possible.

$\frac{49}{50} \div 7 = \frac{49}{50} \div \frac{7}{1} = \frac{49}{50} \times \frac{1}{7} = \frac{49 \times 1}{50 \times 7} = \frac{7}{50}$

$\frac{14}{18} \div 2 = \frac{14}{18} \div \frac{2}{1} = \frac{14}{18} \times \frac{1}{2} = \frac{14 \times 1}{18 \times 2} = \frac{7}{18}$

$\frac{8}{27} \div 18 = \frac{4}{243}$ $\frac{6}{14} \div 21 = \frac{1}{49}$

$\frac{21}{25} \div 14 = \frac{3}{50}$ $\frac{12}{13} \div 24 = \frac{1}{26}$

2. I had originally $30 and spent $\frac{7}{15}$ of it at the café and $\frac{1}{5}$ of it at the movies. How much money did I have left?

$\frac{7}{15} + \frac{1}{5} = \frac{2}{3}$ (have spent)

$30 - \frac{2}{3} \times 30 = 10$

Answer: $10.

Page 114

1. Divide fractions. Write the missing numbers. Simplify if possible.

$\frac{9}{17} \div 18 = \frac{9}{17} \div \frac{18}{1} = \frac{9}{17} \times \frac{1}{18} = \frac{9 \times 1}{17 \times 18} = \frac{1}{34}$

$\frac{24}{35} \div 3 = \frac{24}{35} \div \frac{3}{1} = \frac{24}{35} \times \frac{1}{3} = \frac{24 \times 1}{35 \times 3} = \frac{8}{35}$

$\frac{49}{50} \div x = \frac{7}{50}$ $\frac{15}{16} \div x = \frac{3}{64}$

$x = \frac{49}{50} \div \frac{7}{50} = \frac{49}{50} \times \frac{50}{7} = 7$ $x = \frac{15}{16} \div \frac{3}{64} = \frac{15}{16} \times \frac{64}{3} = 20$

$\frac{36}{41} \div x = \frac{4}{41}$ $\frac{25}{50} \div x = \frac{1}{30}$

$x = \frac{36}{41} \div \frac{4}{41} = \frac{36}{41} \times \frac{41}{4} = 9$ $x = \frac{25}{50} \div \frac{1}{30} = \frac{25}{50} \times \frac{30}{1} = 15$

2. A jar $\frac{5}{6}$ full contains 2 L of water. How much water will the jar contain when $\frac{3}{5}$ full?

$\frac{5}{6} \times x = 2$ ⟶ x = 2.4 (L)

$\frac{3}{5} \times 2.4 = 1.44$ (L)

Answer: 1.44 L.

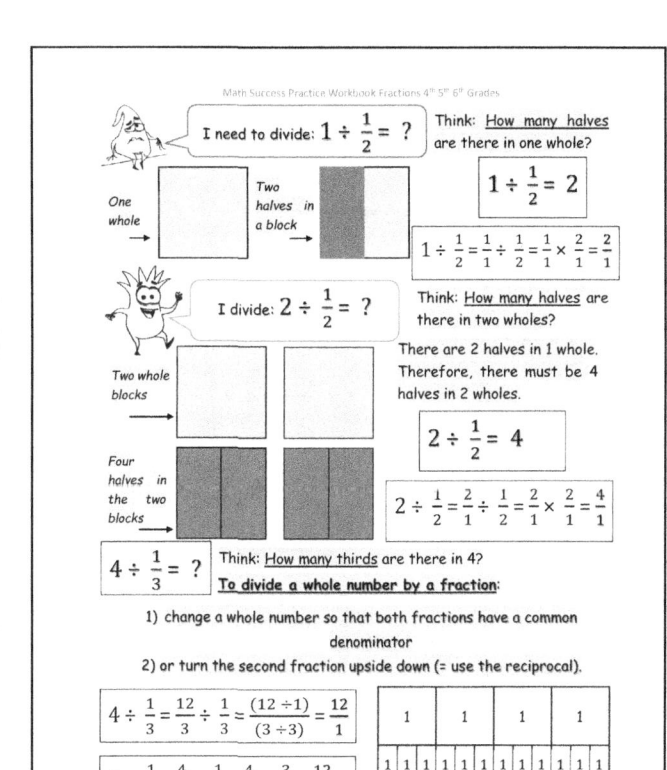

Page 115

I need to divide: $1 \div \frac{1}{2} = ?$ Think: How many halves are there in one whole?

One whole →   Two halves in a block

$1 \div \frac{1}{2} = 2$

$1 \div \frac{1}{2} = \frac{1}{1} \div \frac{1}{2} = \frac{1}{1} \times \frac{2}{1} = \frac{2}{1}$

I divide: $2 \div \frac{1}{2} = ?$ Think: How many halves are there in two wholes?

There are 2 halves in 1 whole. Therefore, there must be 4 halves in 2 wholes.

Two whole blocks

Four halves in the two blocks

$2 \div \frac{1}{2} = 4$

$2 \div \frac{1}{2} = \frac{2}{1} \div \frac{1}{2} = \frac{2}{1} \times \frac{2}{1} = \frac{4}{1}$

$4 \div \frac{1}{3} = ?$ Think: How many thirds are there in 4?

To divide a whole number by a fraction:
1) change a whole number so that both fractions have a common denominator
2) or turn the second fraction upside down (= use the reciprocal).

$4 \div \frac{1}{3} = \frac{12}{3} \div \frac{1}{3} = \frac{(12 \div 1)}{(3 \div 3)} = \frac{12}{1}$

$4 \div \frac{1}{3} = \frac{4}{1} \div \frac{1}{3} = \frac{4}{1} \times \frac{3}{1} = \frac{12}{1}$

Page 116

1. <u>Divide</u> fractions. <u>Write</u> the missing numbers. <u>Simplify</u> if possible.

$8 \div \frac{2}{7} = \frac{8}{1} \div \frac{2}{7} = \frac{8}{1} \times \frac{7}{2} = 28 \text{ ---}$

$6 \div \frac{12}{13} = \frac{6}{1} \div \frac{12}{13} = \frac{6}{1} \times \frac{13}{12} = 6\frac{1}{2}$

$5 \div \frac{10}{17} = \frac{5}{1} \div \frac{10}{17} = \frac{5}{1} \times \frac{17}{10} = 8\frac{1}{2}$

$4 \div \frac{12}{21} = \frac{4}{1} \div \frac{12}{21} = \frac{4}{1} \times \frac{21}{12} = 7 \text{ ---}$

$9 \div \frac{3}{8} = \frac{9}{1} \div \frac{3}{8} = \frac{9}{1} \times \frac{8}{3} = 24 \text{ ---}$

2. I had x points and spent $\frac{3}{4}$ of them on comic books. <u>How many points</u> did I have if I spent 12 points?

$\frac{3}{4}x = 12$

$x = 12 \div \frac{3}{4} = \frac{12}{1} \div \frac{3}{4} = \frac{12}{1} \times \frac{4}{3} = \frac{12 \times 4}{1 \times 3} = 16 \text{ ---}$

Answer: 16 points.

Page 117

1. <u>Divide</u> fractions. <u>Write</u> the missing numbers. <u>Simplify</u> if possible.

$3 \div \frac{21}{25} = \frac{3}{1} \div \frac{21}{25} = \frac{3}{1} \times \frac{25}{21} = 3\frac{4}{7}$

$7 \div \frac{14}{17} = \frac{7}{1} \div \frac{14}{17} = \frac{7}{1} \times \frac{17}{14} = 8\frac{1}{2}$

$3 \div \frac{6}{11} = \frac{3}{1} \div \frac{6}{11} = \frac{3}{1} \times \frac{11}{6} = 5\frac{1}{2}$

$6 \div \frac{24}{25} = 6\frac{1}{4}$ $9 \div \frac{18}{22} = 11$

$3 \div \frac{15}{18} = 3\frac{3}{5}$ $10 \div \frac{10}{15} = 15$

2. Grandma planted x flowers and $\frac{3}{5}$ of them were beautiful 30 lilacs. <u>How many flowers</u> did she plant?

$\frac{3}{5}x = 30$

$x = 30 \div \frac{3}{5} = \frac{30}{1} \div \frac{3}{5} = \frac{30}{1} \times \frac{5}{3} = \frac{30 \times 5}{1 \times 3} = 50 \text{ ---}$

Answer: 50 flowers.

Page 118

1. <u>Divide</u> or <u>multiply</u>. <u>Write</u> the missing numbers. <u>Simplify</u> if possible.

$x \div \frac{5}{7} = 7$ $x \div \frac{4}{15} = 22\frac{1}{2}$

Hint: Change the mixed number to an improper fraction, then divide or multiply!

$x = \frac{7}{1} \times \frac{5}{7} = \frac{7 \times 5}{1 \times 7} = 5$ $x = \frac{45 \times 4}{2 \times 15} = 6$

$x \div \frac{24}{25} = 8\frac{1}{3}$ $x \div \frac{18}{22} = 11$

$x = \frac{25 \times 24}{3 \times 25} = 8$ $x = \frac{11 \times 18}{1 \times 22} = 9$

$x \div \frac{24}{35} = 8\frac{3}{4}$ $x \div \frac{15}{36} = 28\frac{4}{5}$

$x = \frac{35 \times 24}{4 \times 35} = 6$ $x = \frac{144 \times 15}{5 \times 36} = 12$

2. I read x books and $\frac{5}{12}$ of them were 35 comic books. <u>How many books</u> did I read?

$\frac{5}{12}x = 35$

$x = 35 \div \frac{5}{12} = \frac{35}{1} \div \frac{5}{12} = \frac{35}{1} \times \frac{12}{5} = 84$

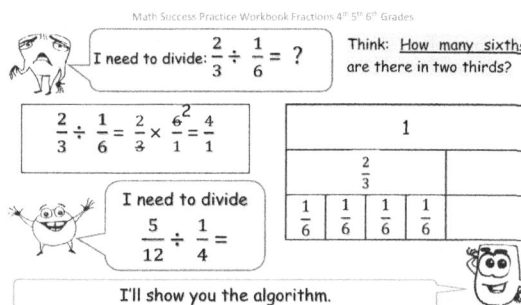

Answer: 84 books.

Page 119

I need to divide: $\frac{2}{3} \div \frac{1}{6} = ?$ Think: How many sixths are there in two thirds?

$\frac{2}{3} \div \frac{1}{6} = \frac{2}{3} \times \frac{6}{1}^{2} = \frac{4}{1}$

I need to divide $\frac{5}{12} \div \frac{1}{4} =$

I'll show you the algorithm.

Step 1: Rewrite the first fraction:

$\frac{5}{12} \div \frac{1}{4} = \frac{5}{12}$

Step 2: Change the operation (÷ → ×):

$\frac{5}{12} \div \frac{1}{4} = \frac{5}{12} \times$

Step 3: Write the reciprocal of the second fraction - switch the numerator and the denominator:

$\frac{5}{12} \div \frac{1}{4} = \frac{5}{12} \times \frac{4}{1}$

Step 4: Multiply the numerators and denominators and simplify where possible:

$\frac{5}{12} \div \frac{1}{4} = \frac{5}{\cancel{12}3} \times \frac{4}{1}$

Step 5: Change to mixed fraction where possible:

$\frac{5}{12} \div \frac{1}{4} = \frac{5}{\cancel{12}3} \times \frac{4}{1} = 6\frac{2}{3}$

Page 120

1. <u>Divide</u> fractions. <u>Write</u> the missing numbers. <u>Simplify</u> if possible.

$\frac{1}{12} \div \frac{5}{6} = \frac{1}{12} \times \frac{6}{5} = \frac{1}{10}$

$\frac{8}{9} \div \frac{24}{45} = \frac{8}{9} \times \frac{45}{24} = 1\frac{2}{3}$

$\frac{5}{18} \div \frac{8}{9} = \frac{5}{18} \times \frac{9}{8} = \frac{5}{16}$

First, change the mixed number to an improper fraction, then, divide or multiply!

$6\frac{5}{6} \div 2\frac{22}{30} = \frac{41}{6} \div \frac{82}{30} = \frac{41}{6} \times \frac{30}{82} = 2\frac{1}{2}$

$1\frac{8}{9} \div 2\frac{14}{27} = \frac{3}{4}$ $\qquad$ $4\frac{3}{7} \div 4\frac{6}{14} = 1$

2. I spent $\frac{1}{3}$ of an hour to find a cave. Then, I spent $\frac{5}{6}$ of an hour to dig out 2 diamonds. Then, I came back to my tiny house, found the recipe of a diamond sword, and spent $\frac{1}{4}$ of an hour to make it. <u>How much time</u> did I spend in all?

$\frac{1}{3} \times 60 = 20 \ (min)$

$\frac{5}{6} \times 60 = 50 \ (min)$

$\frac{1}{4} \times 60 = 15 \ (min)$

Answer: 20 + 50 + 15 = 85 = 1h 25 min.

Page 121

1. <u>Divide</u> fractions. <u>Write</u> the missing numbers. <u>Simplify</u> if possible.

$\frac{14}{18} \div \frac{2}{9} = \frac{14}{18} \times \frac{9}{2} = 3\frac{1}{2}$

$\frac{6}{7} \div \frac{24}{49} = 1\frac{3}{4}$ $\qquad$ $3\frac{2}{6} \div 2\frac{2}{9} = 1\frac{1}{2}$

$\frac{21}{25} \div \frac{14}{15} = \frac{9}{10}$ $\qquad$ $2\frac{5}{8} \div 2\frac{3}{12} = 1\frac{1}{6}$

$1\frac{8}{9} \div 2\frac{2}{33} = \frac{11}{12}$ $\qquad$ $4\frac{2}{7} \div 3\frac{6}{14} = 1\frac{1}{4}$

$8\frac{3}{4} \div 3\frac{6}{8} = 2\frac{1}{3}$ $\qquad$ $10\frac{2}{3} \div 5\frac{4}{6} = 1\frac{15}{17}$

2. A jar is $\frac{3}{10}$ full of water. If 2 gallons are added, then it is $\frac{4}{5}$ full. <u>How many gallons</u> does the jar hold when full?

$\frac{4}{5} - \frac{3}{10} = \frac{1}{2}$ (added)

$\frac{1}{2} x = 2$

$x = 4$

Answer: 4 gallons.

Page 122

1. <u>Divide</u> fractions. <u>Write</u> the missing numbers. <u>Simplify</u> if possible.

$\frac{16}{25} \div \frac{32}{55} = \frac{16}{25} \times \frac{55}{32} = 1\frac{1}{10}$

$\frac{6}{7} \div \frac{24}{49} = \frac{7}{4}$ $\qquad$ $5\frac{1}{7} \div 4\frac{1}{2} = 1\frac{1}{7}$

$\frac{21}{25} \div \frac{14}{15} = \frac{9}{10}$ $\qquad$ $11\frac{3}{7} \div 4\frac{10}{14} = 2\frac{6}{37}$

$15\frac{1}{8} \div 2\frac{1}{5} = 6\frac{7}{8}$ $\qquad$ $4\frac{1}{5} \div 2\frac{1}{3} = 1\frac{4}{5}$

$x \div 3\frac{4}{8} = 2\frac{2}{7}$ $\qquad$ $x \div 7\frac{1}{7} = 3\frac{9}{25}$

$x = 8$ $\qquad$ $x = 24$

2. I save $\frac{7}{9}$ of my weekly pocket money. Of the remainder, I spend $\frac{3}{5}$ of it on movies. <u>What fraction of the total money</u> do I spend on the movies?

$\frac{9}{9} - \frac{7}{9} = \frac{2}{9}$ (the remainder)

$\frac{3}{5} \times \frac{2}{9} = \frac{2}{15}$

Answer: $\frac{2}{15}$.

Made in the USA
Columbia, SC
20 September 2025